Contents

Skills Practice

Using a Problem-Solving Plan

-1

Solve each problem.

1. **MONEY** Jackie wants to take out an ad in the newspaper for her up-coming garage sale. She can buy a 4-line ad for $4.35 that will run for three days. If she wants to spend no more than $15 on advertising, how long can she advertise?

2. **BASEBALL** Cy Young pitched in 815 games over 22 years. He won 511 games. About how many games did he win per year?

3. **MONEY** Each month, Kevin's car costs $59 for insurance, $42 for maintenance, and $58 for gas. About how much does it cost Kevin to drive his car for a year?

4. **MONEY** How many ways can you make change for a dollar using nickels, dimes, and/or quarters?

5. **FOOD** Friday night Joe decided to order a 1-topping pizza. He had a choice of thin or thick crust and a choice of five toppings (pepperoni, mushrooms, sausage, onions, or peppers). How many different pizzas could he choose from?

Find the next term in each list.

6. 7, 11, 15, 19, 23, ...

7. 2, 4, 8, 16, 32, ...

8. 63, 54, 45, 36, 27, ...

9. 3, 0, 5, 3, 0, 5, ...

10. 0, 4, 8, 12, 16, ...

GEOMETRY **Draw the next figure in each pattern.**

11. □ ▲ □ □ ▲ ▲ □ □ □ ...

12. ... ...

1-2 Skills Practice

Numbers and Expressions

Name the operation that should be performed first. Then find the value of each expression.

1. $2 - 3 \cdot 0$

2. $25 \div 5 - 4$

3. $5 + 2 - 3$

4. $2 \cdot 5 + 6$

5. $9 \div 3 \cdot 2 + 1$

6. $5 + 2 \cdot 8 + 2 - 5$

Find the value of each expression.

7. $4 + 2 \cdot 8$

8. $30 - 12 \cdot 2$

9. $5 + 2 \cdot 3 + 4$

10. $10 - 2 \cdot 4 - 1$

11. $15 - 10 \div 2$

12. $25 - 6 \cdot 4 + 9$

13. $(14 + 6) \div 5$

14. $100 + 50 \div 10$

15. $14 - (4 \cdot 2)$

16. $(3 + 4) \cdot (5 + 3)$

17. $6(4 + 5)$

18. $\dfrac{(8 \cdot 9)}{(3 \cdot 4)}$

19. $(2 + 3) \cdot 5 + 1$

20. $24 - 24 \div 8$

21. $56 \div (3 + 4)$

22. $2[(4 + 5) \cdot 3]$

Write a numerical expression for each verbal phrase.

23. the difference of seventeen and three

24. eleven more than six

25. the sum of eight, twenty, and thirty-five

26. the quotient of forty and eight

27. one hundred decreased by twenty-five

28. three more than one dozen

29. the product of twenty and thirty

30. five less than fifty

Skills Practice

Variables and Expressions

ALGEBRA Evaluate each expression if $x = 4$, $y = 6$, and $z = 3$.

1. $x + y + z$
2. $3x + y$
3. $x - z$
4. $x + y - 3z$
5. $15z$
6. $3(x + z)$
7. $xz \div y$
8. $yz - x$

ALGEBRA Evaluate each expression if $a = 7$, $b = 9$, $c = 2$, and $d = 5$.

9. $a + b + c$
10. $a + b - (c + d)$
11. $3a + 4d$
12. bcd
13. $(a + b) \cdot (c + d)$
14. $c(4 + d)$
15. $\dfrac{b}{a + c}$
16. $a + b - 3c$
17. $ab - cd$
18. $\dfrac{bc}{a - d}$

ALGEBRA Translate each phrase into an algebraic expression.

19. two inches shorter than Kathryn's height

20. the quotient of some number and thirteen

21. some number added to seventeen

22. six centimeters shorter than the length of the pencil

23. three pounds lighter than Adlai's weight

24. the difference of some number and eighteen

25. three dollars more than the cost of a ticket

26. eight more than the product of a number and four

27. half as many pieces of candy

28. twice as long as the length of the string

1-4 **Skills Practice**

Properties

Name the property shown by each statement.

1. $9 \cdot 7 = 7 \cdot 9$

2. $37 \cdot 0 = 0$

3. $1 \cdot 87 = 87$

4. $14 + 6 = 6 + 14$

5. $3(6a) = (3 \cdot 6)a$

6. $2b + 0 = 2b$

Find each sum or product mentally.

7. $4 + 23 + 46$

9. $2 \cdot 15 \cdot 10$

11. $14 + 24 + 6 + 26$

13. $25 \cdot 0 \cdot 8$

8. $327 \cdot 6 \cdot 0$

10. $5 \cdot 16 \cdot 20$

12. $43 + 38 + 7$

14. $11 + 28 + 19$

ALGEBRA Simplify each expression.

15. $(x + 5) + 4$

17. $38 + (v + 12)$

19. $16p \cdot 0$

21. $8(p9)$

23. $3(11k)$

25. $m(13 \cdot 5)$

16. $(6a)10$

18. $8(q3)$

20. $16 + (22 + x)$

22. $(17 + 33) + x$

24. $16 + (y + 9)$

26. $17 + (n + 0)$

1-5 Skills Practice

Variables and Equations

ALGEBRA Find the solution of each equation from the list given.

1. $u + 11 = 42$; 29, 31, 33

2. $23 + w = 30$; 7, 8, 9

3. $18 + 17 = g$; 33, 34, 35

4. $s - 16 = 4$; 18, 20, 22

5. $17 - x = 2$; 13, 15, 17

6. $27 - 6 = d$; 17, 19, 21

7. $8r = 24$; 3, 4, 5

8. $16 = 4v$; 2, 3, 4

9. $\dfrac{42}{x} = 7$; 6, 8, 10

10. $\dfrac{x}{11} = 7$; 73, 75, 77

ALGEBRA Solve each equation mentally.

11. $c + 9 = 11$

12. $20 = f + 5$

13. $m - 6 = 3$

14. $24 - u = 18$

15. $6r = 36$

16. $8h = 40$

17. $p + 9 = 25$

18. $6 = \dfrac{30}{z}$

ALGEBRA Define a variable. Then write an equation and solve.

19. The product of 5 and a number is 50.

20. A number decreased by 3 is 9.

21. Eleven more than a number is 33.

22. Forty divided by a number is 8.

23. A number times 9 is 36.

24. A number divided by 7 is 7.

25. Thirty less than 40 is a number.

26. The difference between 20 and 8 is a number.

27. The sum of 4 and a number is 20.

28. The quotient of 50 and a number is 2.

1-6 Skills Practice

Ordered Pairs and Relations

Graph each point on the coordinate system.

1. $A(2, 5)$ **2.** $M(6, 4)$

3. $Z(1, 1)$ **4.** $R(3, 0)$

5. $Q(7, 8)$ **6.** $W(0, 6)$

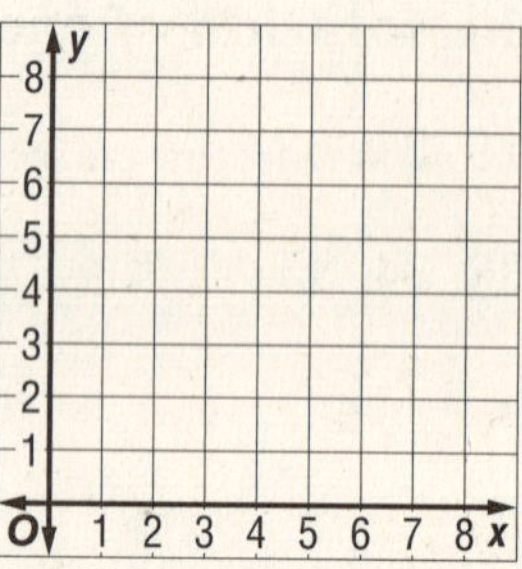

Write the ordered pair that names each point.

7. N **8.** K

9. A **10.** V

11. Z **12.** G

13. R **14.** B

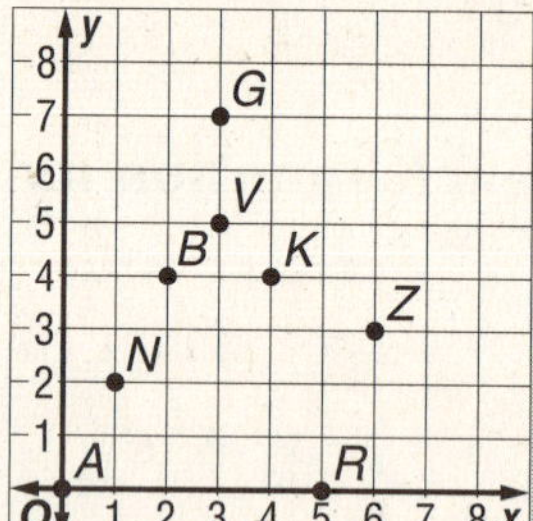

Express each relation as a table and as a graph. Then determine the domain and range.

15. $\{(3, 7), (1, 1), (6, 5), (2, 4)\}$

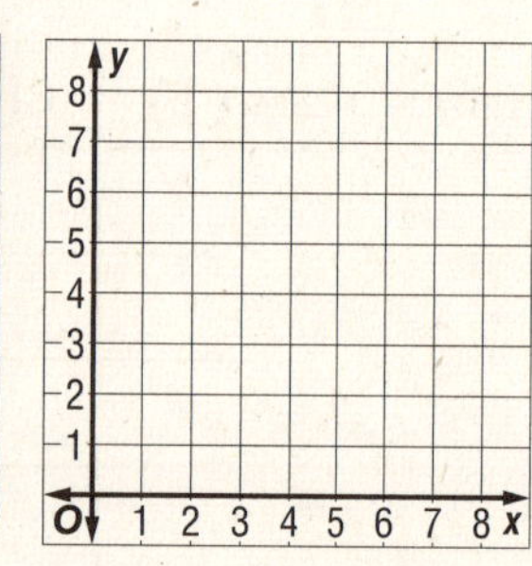

16. $\{(0, 3), (5, 7), (1, 8)\}$

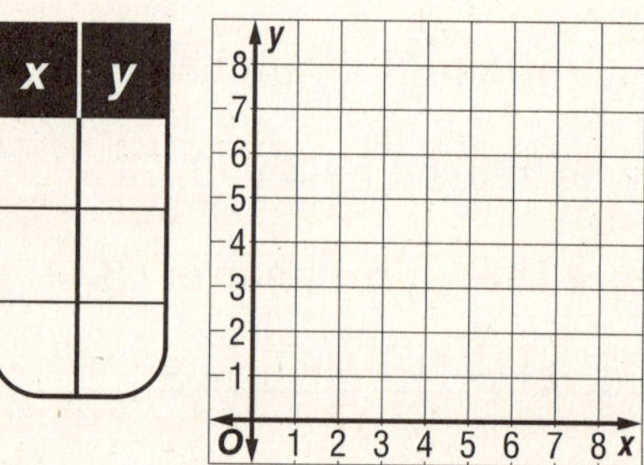

17. $\{(2, 3), (3, 2), (1, 7), (7, 1)\}$

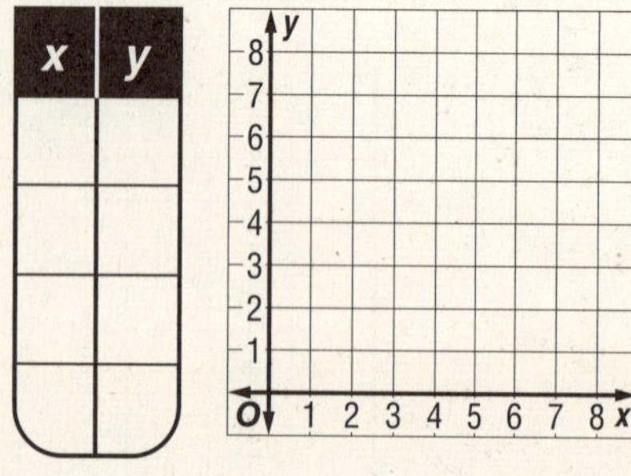

18. $\{(5, 6), (0, 2), (4, 4) (8, 3)\}$

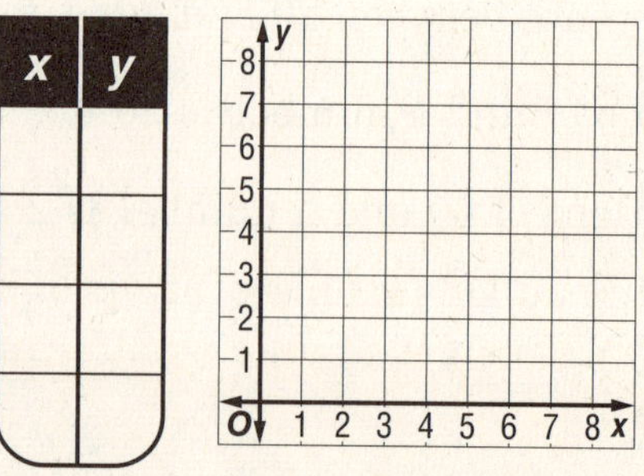

1-7 Skills Practice

Scatter Plots

Tell whether each scatter plot shows a *positive*, *negative*, or *no* relationship.

1. 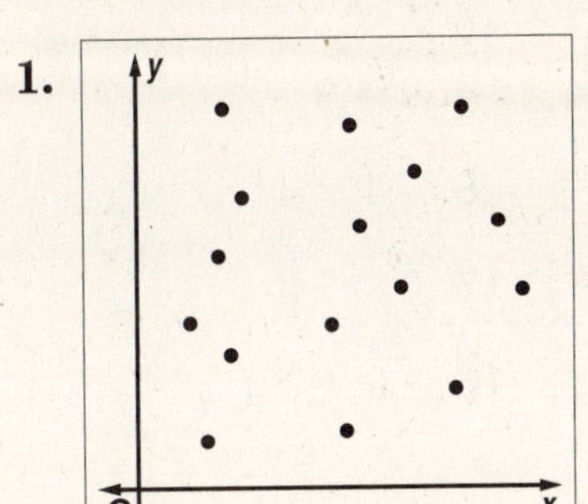

2. 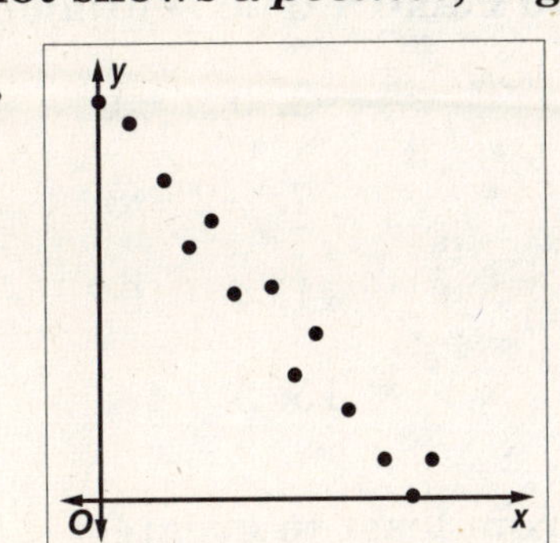

3. 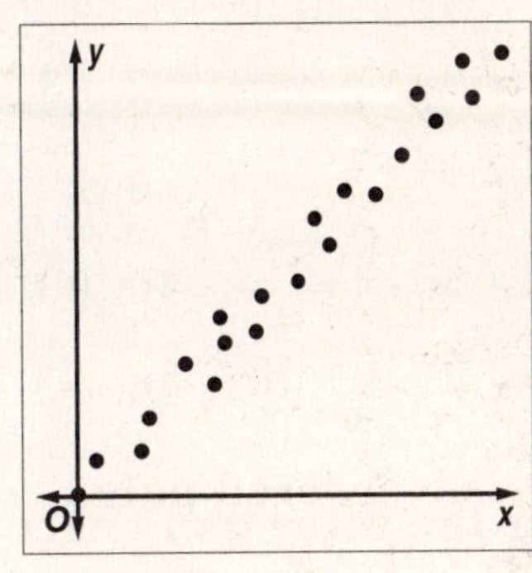

4. Draw a scatter plot with six ordered pairs that shows a positive relationship. Explain your reasoning.

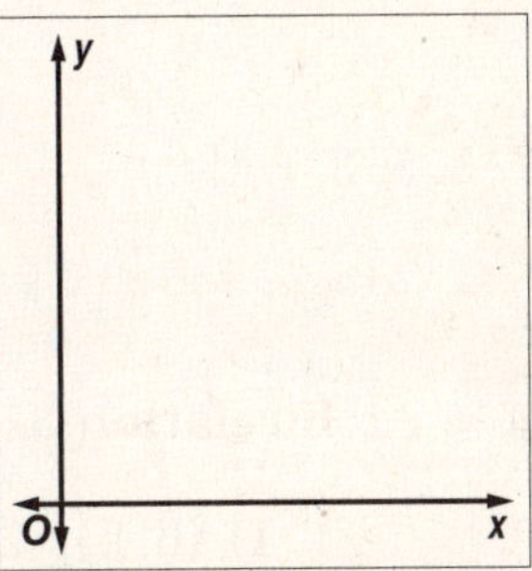

For Exercises 5–8, use the following information:

SCIENCE Scientists have determined that there may be a relationship between temperature and the number of chirps produced by crickets. The table gives the temperature and the number of chirps per minute for several cricket samples.

Temperature (°F)	Chirps/min
71	138
68	97
75	152
80	158
60	81
75	155
84	165

5. Make a scatter plot of the data.

6. Does there appear to be a relationship between temperature and chirps? Explain.

7. Suppose the outside temperature is 65°. About how many chirps per minute would you expect from a cricket?

8. Suppose the outside temperature is 55°. About how many chirps per minute would you expect from a cricket?

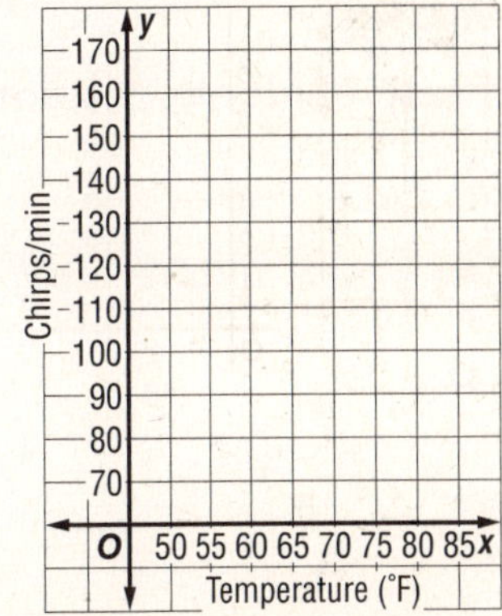

2-1 Skills Practice

Integers and Absolute Value

Replace each ● with <, >, or = to make a true sentence.

1. 1 ● 0 **2.** −3 ● 0 **3.** 0 ● −1 **4.** 0 ● 9

5. −7 ● −7 **6.** 2 ● −2 **7.** −2 ● 8 **8.** −4 ● 4

9. 5 ● 5 **10.** 0 ● −6 **11.** 4 ● 10 **12.** 6 ● −6

13. 3 ● 7 **14.** −1 ● −2 **15.** 3 ● 4 **16.** −3 ● −4

Order the integers in each set from least to greatest.

17. {4, −5, 0} **18.** {8, −2, 1} **19.** {−6, −3, 0}

20. {−5, 5, 3, −1} **21.** {0, −3, 7, −2} **22.** {9, −11, 1, 0}

23. {12, −4, 3, −1} **24.** {−8, 15, 1, −10} **25.** {−12, −17, −20, 2}

Evaluate each expression.

26. $|1|$ **27.** $|-10|$ **28.** $|-8|$

29. $|10|$ **30.** $|4| + |-4|$ **31.** $|9| - |-5|$

32. $0 + |-1|$ **33.** $|-6| + |-5|$ **34.** $|-8| - |-8|$

35. $|12| + |-3|$ **36.** $|-15| - |6|$ **37.** $|-13| + |-7|$

Evaluate each expression if $a = -3$, $b = 0$, and $c = 1$.

38. $|a| - b$ **39.** $|c| + 2$ **40.** $9 - |a|$

41. $|25| - b$ **42.** $10 - |b|$ **43.** $|-8| + |a|$

Glencoe Pre-Algebra

2-2 Skills Practice

Adding Integers

Find each sum.

1. $-7 + (-5)$ **2.** $10 + 9$ **3.** $-12 + (-5)$ **4.** $-13 + (-3)$

5. $-10 + 12$ **6.** $-7 + 8$ **7.** $-11 + (-6)$ **8.** $0 + (-21)$

9. $72 + (-10)$ **10.** $72 + 10$ **11.** $-13 + (-11)$ **12.** $-52 + 52$

13. $-6 + (-12)$ **14.** $14 + (-8)$ **15.** $-17 + (-2)$ **16.** $50 + (-8)$

17. $-22 + 4$ **18.** $14 + 8$ **19.** $-21 + (-9)$ **20.** $15 + (-5)$

21. $9 + 10$ **22.** $-12 + (-15)$ **23.** $-13 + 6$ **24.** $-1 + (-18)$

25. $0 + 31$ **26.** $-45 + (-15)$ **27.** $-6 + 20$ **28.** $24 + (-11)$

29. $7 + (-14)$ **30.** $-34 + (-10)$ **31.** $-8 + (-25)$ **32.** $-31 + 25$

33. $6 + 5 + (-4)$ **34.** $-4 + (-5) + 6$ **35.** $-3 + 8 + (-9)$

36. $-6 + (-2) + (-1)$ **37.** $10 + (-5) + 6$ **38.** $-8 + 8 + (-10)$

39. $0 + (-8) + 22$ **40.** $-31 + 19 + (-15)$ **41.** $32 + (-4) + (-9)$

2-3 Skills Practice

Subtracting Integers

Find each difference.

1. $-2 - (-8)$ **2.** $4 - (-11)$ **3.** $-7 - 6$ **4.** $15 - 2$

5. $-7 - (-1)$ **6.** $1 - 9$ **7.** $-5 - (-3)$ **8.** $6 - (-5)$

9. $-4 - (-10)$ **10.** $4 - 6$ **11.** $0 - (-15)$ **12.** $-16 - (-10)$

13. $0 - 16$ **14.** $11 - (-9)$ **15.** $-9 - 1$ **16.** $-1 - (-8)$

17. $1 - (-2)$ **18.** $-2 - (-19)$ **19.** $13 - 17$ **20.** $20 - (-15)$

21. $-10 - (-21)$ **22.** $4 - 22$ **23.** $-8 - 16$ **24.** $12 - (-9)$

Evaluate each expression if $a = -9$, $b = 4$, and $c = -5$.

25. $a - 8$ **26.** $10 - c$ **27.** $11 - b$ **28.** $15 - a$

29. $-8 - b$ **30.** $c - 1$ **31.** $-32 - a$ **32.** $b - 25$

33. $c - (-14)$ **34.** $-33 - a$ **35.** $14 - c$ **36.** $b - c$

37. $a - c$ **38.** $b - a$ **39.** $c - b$ **40.** $c - a$

41. $a - b$ **42.** $a + b - c$ **43.** $b + 15 + a$ **44.** $a - (-b) + c$

Glencoe Pre-Algebra

2-4 Skills Practice

Multiplying Integers

Find each product.

1. $-2(8)$ **2.** $-4(-4)$ **3.** $6(-2)$ **4.** $-7(-3)$

5. $12(1)$ **6.** $0(-2)$ **7.** $-11(5)$ **8.** $-9(-3)$

9. $-13(0)$ **10.** $-1(-1)$ **11.** $-1(1)$ **12.** $1(-1)$

13. $-5(20)$ **14.** $16(-2)$ **15.** $18(-3)$ **16.** $-5(-5)$

17. $8(6)(-2)$ **18.** $-1(50)(-1)$ **19.** $6(0)(-2)$ **20.** $(-3)(-2)(-1)$

21. $-4(5)(-3)$ **22.** $10(-3)(2)$ **23.** $-9(8)(1)$ **24.** $-1(-1)(-1)$

ALGEBRA Simplify each expression.

25. $-2 \cdot 3x$ **26.** $-4 \cdot 5y$ **27.** $9 \cdot (-2z)$ **28.** $-5 \cdot (-6a)$

29. $8t \cdot (-3)$ **30.** $2n \cdot (-1)$ **31.** $-5 \cdot 2w$ **32.** $8c \cdot (-2)$

33. $-3c \cdot (-5d)$ **34.** $4r \cdot 7s$ **35.** $-3x \cdot (-z)$ **36.** $-4ab \cdot (-6)$

37. $(-3)(4)(-x)$ **38.** $-3(5)(-y)$ **39.** $(-6)(-2)(8r)$ **40.** $-5(0)(-xy)$

ALGEBRA Evaluate each expression if $x = -5$ and $y = -6$.

41. $3y$ **42.** $-8x$ **43.** $-4y$ **44.** $12x$

45. xy **46.** $-xy$ **47.** $-6xy$ **48.** $4xy$

2-5 Skills Practice

Dividing Integers

Find each quotient.

1. $16 \div 4$ **2.** $-27 \div 3$ **3.** $25 \div (-5)$ **4.** $63 \div (-9)$

5. $-15 \div (-3)$ **6.** $14 \div (-7)$ **7.** $-124 \div 4$ **8.** $60 \div 15$

9. $28 \div (-4)$ **10.** $-56 \div (-8)$ **11.** $72 \div 8$ **12.** $-21 \div (-7)$

13. $\dfrac{-32}{4}$ **14.** $\dfrac{45}{9}$ **15.** $\dfrac{-45}{3}$ **16.** $\dfrac{-25}{-5}$

17. $\dfrac{35}{-7}$ **18.** $\dfrac{-63}{-7}$ **19.** $\dfrac{-144}{12}$ **20.** $\dfrac{48}{-6}$

21. What is -54 divided by 9?   **22.** Divide -27 by -3.

23. Divide 144 by -12.   **24.** What is -65 divided by -13?

Evaluate each expression if $x = -8$ and $y = -12$.

25. $x \div 2$ **26.** $x \div (-4)$ **27.** $36 \div y$ **28.** $0 \div y$

29. $-60 \div y$ **30.** $56 \div x$ **31.** $8 \div x$ **32.** $-108 \div y$

33. $\dfrac{x}{-2}$ **34.** $\dfrac{y}{3}$ **35.** $\dfrac{0}{x}$ **36.** $\dfrac{-112}{x}$

37. $\dfrac{y}{-6}$ **38.** $\dfrac{x}{4}$ **39.** $\dfrac{-144}{y}$ **40.** $\dfrac{-136}{x}$

Find the average (mean) of each group of numbers.

41. $3, 12, 6$ **42.** $-8, -1, -3$ **43.** $-8, 15, 5, 8$ **44.** $-3, -10, 2, -4, 0$

45. $-10, -7, 7, 10$ **46.** $12, 24, 9, 15, 18, 20, 16, 14$ **47.** $-4, -11, -6, 1, 8, -12$

Glencoe Pre-Algebra

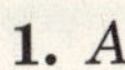 **2-6**

Skills Practice

The Coordinate System

Name the ordered pair for each point graphed at the right.

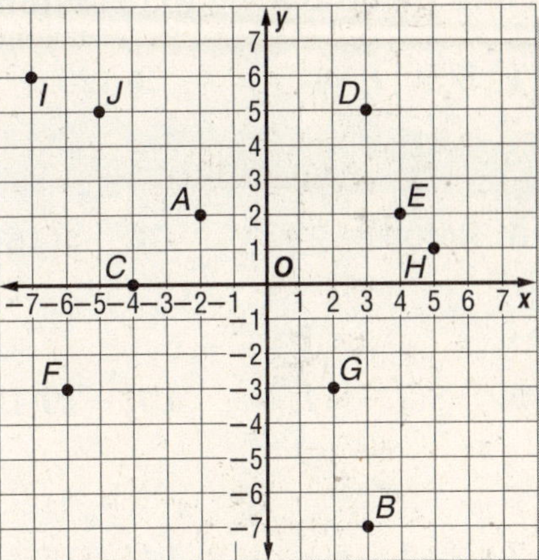

1. *A*

2. *B*

3. *C*

4. *D*

5. *E*

6. *F*

7. *G*

8. *H*

9. *I*

10. *J*

Graph and label each point on the coordinate plane. Name the quadrant in which each point is located.

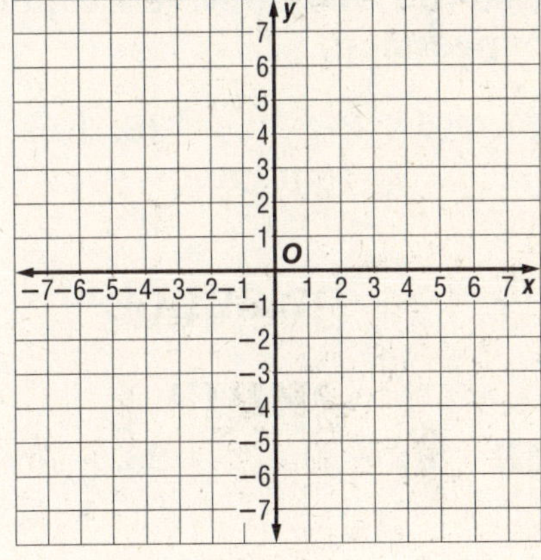

11. $K\,(1, 0)$

12. $L\,(0, 2)$

13. $M\,(-2, 4)$

14. $N\,(-5, -4)$

15. $P\,(6, -2)$

16. $Q\,(7, -6)$

17. $R\,(-3, -4)$

18. $S\,(4, -7)$

19. $T\,(3, 6)$

20. $U\,(-7, 4)$

21. **ALGEBRA** Make a table of values and graph six sets of ordered pairs for the equation $y = x - 4$. Describe the graph.

$y = x - 4$		
x	**y**	**(x, y)**

3-1 **Skills Practice**

The Distributive Property

Use the Distributive Property to write each expression as an equivalent expression. Then evaluate the expression.

1. $8(50 + 4)$

2. $(20 + 9)5$

3. $2(60 + 4)$

4. $7(40 - 2)$

5. $4(400 - 2)$

6. $-4(16 + 5)$

7. $-8(4 + 1)$

8. $9(24 - 19)$

9. $-3(7 - 11)$

10. $-10(12 - 4)$

11. $(21 + 9)(-5)$

12. $-7(1 - 10)$

13. $-2(1 - 6)$

14. $4(15 + 25)$

15. $15(100 + 6)$

16. $12(22 - 52)$

Use the Distributive Property to write each expression as an equivalent algebraic expression.

17. $4(d + 2)$

18. $1(u - 3)$

19. $-6(f + 5)$

20. $-2(g - 3)$

21. $3(x - 7)$

22. $8(-b + 4)$

23. $(9 - h)5$

24. $(c + 1)(-4)$

25. $-1(2 - y)$

26. $-7(a + 1)$

27. $11(k - 20)$

28. $-9(r - 1)$

29. $5(1 - b)$

30. $8(x + 12)$

31. $-6(p + 15)$

32. $4(h - 16)$

33. $-3(w - 10)$

34. $-10(c + 9)$

35. $2(11 - q)$

36. $-4(12 - f)$

37. $12(n + 2)$

38. $16(g + 1)$

39. $-8(9 + b)$

40. $-5(z - 4)$

41. $6(r - 20)$

42. $7(2 - j)$

43. $-1(m + 1)$

44. $-2(v - 8)$

45. $5(q - 16)$

46. $-10(c - 7)$

47. $-3(-x - 1)$

48. $(9 - h)(-2)$

3-2 Skills Practice

Simplifying Algebraic Expressions

Simplify each expression.

1. $7a + a$

2. $k - k$

3. $m + 3m + 8$

4. $10b - b + 1$

5. $9j + 8j - 7j$

6. $6y + 3y + 6y - 2y$

7. $3q + 2q - q$

8. $18 + 7x - 12 + 5x$

9. $12a + 3 + 18 - 9a$

10. $13c - 7 + c - d$

11. $5h + h - 4h + 1 - 2h$

12. $2(v - 5) + 7v + 4$

13. $5(r + 9) - 5$

14. $1 - 4(u - 1)$

15. $-7(w - 4) + 3w - 27$

16. $-8 - 7(y + 2)$

17. $-18(c - 1) - 18$

18. $12(n - 4) - 3n$

19. $5m - 9 + 4m$

20. $-7 + g + 1 - 6g$

21. $x - 9x + 3 + 8x - 3$

22. $6(r - 4) + r + 30 - 7r$

23. $-5 + 5a - 4 - 2a + 3a$

24. $21 - 8(v + 3) + 3 + 7v$

25. $4x - 9 + 3x + 6 - 9x - 4$

26. $p - 2 + 1 - p + 1 + 2p$

27. $-11f + 6 - f + 4 + 13f - 9$

28. $3(d - 4) + 2 - 2d + 1 - d$

29. $1 - s + 2 + 2s - 3s + 1$

30. $5 - 9k + 1 + k - 2(7 - k)$

31. $1 - g + 5 - 2g + 3(g - 2)$

32. $7h + 1 - h + 4 - 2 - 8h$

33. $-12 + 7(d - 1) + 14 - d$

3-3 Skills Practice

Solving Equations by Adding or Subtracting

Solve each equation. Check your solution.

1. $r + 1 = -5$ **2.** $h + 8 = 6$ **3.** $t - 3 = -11$ **4.** $p - 5 = 9$

5. $w + 9 = -9$ **6.** $x - 9 = -9$ **7.** $a + 7 = -7$ **8.** $m + 9 = -7$

9. $q - 4 = 5$ **10.** $b + 2 = 3$ **11.** $n - 11 = 1$ **12.** $r - 1 = -3$

13. $c + 6 = 1$ **14.** $v - 3 = -7$ **15.** $z + 3 = 0$ **16.** $s - 8 = -1$

17. $y - 7 = -5$ **18.** $u - 10 = -2$ **19.** $g + 1 = 10$ **20.** $k + 4 = -9$

21. $w + 12 = -4$ **22.** $z - 8 = -8$ **23.** $d - 11 = 1$ **24.** $h + 3 = 10$

25. $r + 10 = -6$ **26.** $y + 1 = 4$ **27.** $f - 6 = 6$ **28.** $d - 2 = -8$

29. $j + 11 = 4$ **30.** $m - 10 = 4$ **31.** $q + 3 = -5$ **32.** $g - 4 = 0$

33. $a - 12 = -19$ **34.** $c + 5 = 2$ **35.** $h - 9 = 12$ **36.** $p + 14 = -1$

37. $v + 13 = -11$ **38.** $x + 8 = -1$ **39.** $y + 12 = -10$ **40.** $k - 16 = 7$

41. $d - 15 = -14$ **42.** $g - 12 = 10$ **43.** $b + 13 = -20$ **44.** $f - 15 = -1$

45. $q + 8 = 13$ **46.** $w - 4 = -15$ **47.** $r + 10 = -13$ **48.** $t - 11 = 11$

49. $j - 9 = -8$ **50.** $k + 2 = -15$ **51.** $n + 12 = 0$ **52.** $y + 9 = 14$

3-4 Skills Practice

Solving Equations by Multiplying or Dividing

Solve each equation. Check your solution.

1. $3x = 24$

2. $\dfrac{m}{-5} = -15$

3. $-4f = 16$

4. $\dfrac{u}{2} = 12$

5. $-6a = 6$

6. $\dfrac{s}{-1} = 10$

7. $-2y = -2$

8. $-7z = 7$

9. $\dfrac{n}{8} = -24$

10. $-4r = -12$

11. $-9h = 81$

12. $\dfrac{c}{-10} = 1$

13. $\dfrac{v}{-15} = -15$

14. $\dfrac{m}{12} = 0$

15. $-12g = 12$

16. $\dfrac{w}{-4} = 0$

17. $-1f = 11$

18. $\dfrac{r}{-1} = 22$

19. $8d = -16$

20. $\dfrac{r}{15} = 45$

21. $25k = -200$

22. $-3p = 18$

23. $7j = -63$

24. $\dfrac{y}{-10} = 10$

25. $\dfrac{x}{-8} = -1$

26. $5g = -20$

27. $\dfrac{p}{6} = 0$

28. $7y = 7$

29. $-6q = -30$

30. $-12c = -60$

31. $-9b = 90$

32. $-4k = -120$

33. $2r = 0$

34. $-1t = 19$

35. $\dfrac{n}{-12} = 12$

36. $-15j = 120$

37. $\dfrac{u}{-11} = 11$

38. $5c = 85$

39. $-9q = -36$

40. $9z = -144$

3-5   **Skills Practice**

Solving Two-Step Equations

Solve each equation. Check your solution.

1. $3x + 10 = 1$

2. $\dfrac{a}{5} + 8 = 9$

3. $8w - 12 = -4$

4. $\dfrac{r}{2} + 6 = 5$

5. $18 - 2q = 4$

6. $3j - 20 = 16$

7. $\dfrac{u}{12} - 8 = -8$

8. $7p + 11 = -31$

9. $12d + 15 = 3$

10. $4c + 20 = 0$

11. $\dfrac{n}{2} - 9 = -5$

12. $10b - 19 = 11$

13. $2h + 10 = -12$

14. $6k - 9 = 15$

15. $\dfrac{w}{-5} - 4 = -2$

16. $12 - 7y = -2$

17. $11 - 3g = 32$

18. $12s + 13 = 25$

19. $2z - 4 - z = 4$

20. $10 - 5h + 2 = 32$

21. $\dfrac{r}{-7} - 5 = -6$

22. $-4a + 5 - 2a - 9 = 44$

23. $\dfrac{w}{-3} + 6 - 1 = 2$

24. $7k - 8k = 1$

25. $7f - 24 = 25$

26. $6 - \dfrac{m}{6} - 8 = 0$

27. $10 - d = 19$

28. $9x + 5 - 4x = -20$

29. $3 - 4t + 11 = 2$

30. $\dfrac{a}{3} - 4 + 9 = 7$

31. $6q - 4 = -16$

32. $\dfrac{m}{8} - 12 - 3 = -12$

33. $5b + 6 - 6b + 2 = 19$

3-6 Skills Practice

Writing Two-Step Equations

Translate each sentence into an equation. Then find each number.

1. Eleven less than 5 times a number is 24.

2. The quotient of a number and -9 increased by 10 is 11.

3. Five less than the product of -3 and a number is -2.

4. Fifteen more than twice a number is -23.

5. The difference between 5 times a number and 4 is 16.

6. Nine more than -8 times a number is -7.

7. The difference between 12 and ten times a number is -28.

8. Seven more than three times a number is 52.

9. Eleven less than five times a number is 19.

10. Thirteen more than four times a number is -91.

11. Seven less than twice a number is 43.

Solve each problem by writing and solving an equation.

12. **SHOPPING** The total cost of a suit and 4 ties is $292. The suit cost $200. Each tie cost the same amount. Find the cost of one tie.

13. **AGES** Mary's sister is 7 years older than Mary. Their combined ages add up to 35. How old is Mary?

3-7 Skills Practice

Using Formulas

1. AIR TRAVEL   A plane is traveling 9 miles per minute. How much time is needed to travel 216 miles?

2. JOGGING   What is the rate, in feet per second, of a girl who jogs 315 feet in 45 seconds?

Find the perimeter and area of each rectangle.

3.

2 mm

6 mm

4.

5 ft

4 ft

5.

45 m

75 m

6.

84 mi

126 mi

7. a rectangle that is 21 inches long and 13 inches wide

8. a square that is 25 centimeters on each side

Find the missing dimension of each rectangle.

9.

20 in.

Perimeter = 208 in.

10. 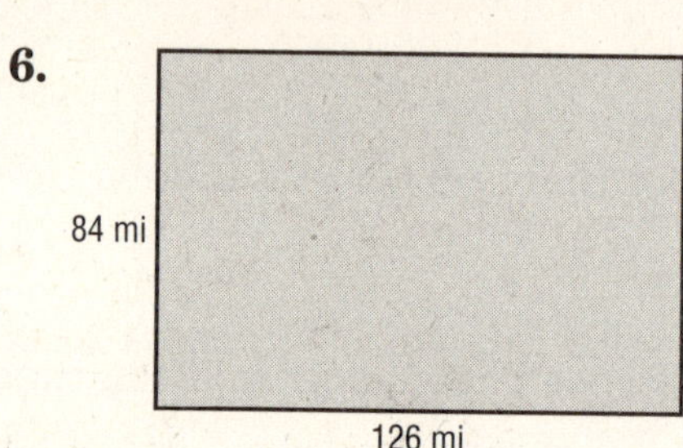

11.

625 mi

Perimeter = 2402 mi

12. 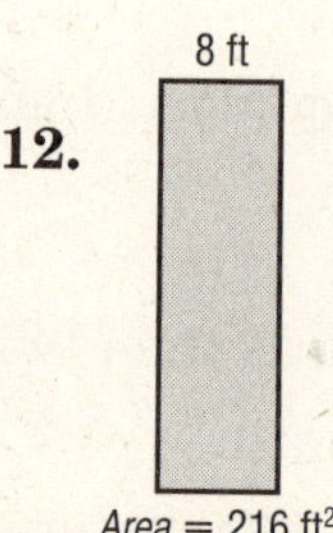

13. The perimeter of a rectangle is 100 centimeters. Its width is 9 centimeters. Find its length.

14. The area of a rectangle is 319 square kilometers. Its width is 11 kilometers. Find its length.

4-1 Skills Practice

Factors and Monomials

Use divisibility rules to determine whether each number is divisible by 2, 3, 5, 6, or 10.

1. 100 **2.** 66

3. 88 **4.** 123

5. 240 **6.** 280

7. 255 **8.** 165

9. 318 **10.** 1000

List all the factors of each number.

11. 36 **12.** 29

13. 45 **14.** 81

15. 125 **16.** 117

17. 16 **18.** 63

Determine whether each expression is a monomial. Explain why or why not.

19. p **20.** 73

21. $2 + n$ **22.** $h - w$

23. $3(a + 6)$ **24.** $-3k$

25. $q + r$ **26.** $4y - 6$

27. $3(x - 3)$ **28.** $6s \cdot 4p$

29. SEATING Can 132 graduates be seated in rows of 6 at the graduation ceremony? Explain.

30. SCHOOL SUPPLIES When Alex's mother buys pencils for school, she divides them equally among Alex and his sister. Should she buy the pencils in packages of 15 or 30? Explain.

4-2 Skills Practice

Powers and Exponents

Write each expression using exponents.

1. $7 \cdot 7$

2. $(-3)(-3)(-3)(-3)(-3)$

3. 4

4. $(k \cdot k)(k \cdot k)(k \cdot k)$

5. $p \cdot p \cdot p \cdot p \cdot p \cdot p$

6. $3 \cdot 3$

7. $(-a)(-a)(-a)(-a)$

8. $6 \cdot 6 \cdot 6 \cdot 6$

9. $9 \cdot 9 \cdot 9$

10. $4 \cdot y \cdot z \cdot z \cdot z$

11. $s \cdot s \cdot s \cdot s \cdot t \cdot u \cdot u$

12. $5 \cdot 5 \cdot 5 \cdot q \cdot q$

Express each number in expanded form.

13. 135

14. 8732

15. 1005

16. 989

Evaluate each expression if $b = 8$, $c = 2$, and $d = -3$.

17. 4^c

18. c^0

19. b^3

20. $c^3 \cdot 3^c$

21. 3^c

22. c^4

23. $c^2 + d$

24. $2b^2$

25. $b^2 + c^3$

26. d^2

27. d^3

28. $b^2 + d^3$

29. $b^2 d$

30. $(b - c)^2$

4-3 Skills Practice

Prime Factorization

Determine whether each number is *prime* or *composite*.

1. 41

2. 29

3. 87

4. 36

5. 57

6. 61

7. 71

8. 103

9. 39

10. 91

11. 47

12. 67

Write the prime factorization of each number. Use exponents for repeated factors.

13. 20

14. 40

15. 32

16. 44

17. 90

18. 121

19. 46

20. 30

21. 65

22. 80

Factor each monomial.

23. $15t$

24. $16r^2$

25. $-11m^2$

26. $-49y^3$

27. $21ab$

28. $-42xyz$

29. $45j^2k$

30. $17u^2v^2$

31. $27d^4$

32. $-16cd^2$

4-4 Skills Practice

Greatest Common Factor (GCF)

Find the GCF of each set of numbers or monomials.

1. 15, 50
2. 24, 81
3. 18, 27
4. 36, 64
5. 88, 40
6. 54, 63
7. 11, 22
8. 14, 25
9. 20, 30
10. 16, 18
11. 64, 80
12. 16, 24
13. $30t, 40t^2$
14. $6, 9t$
15. $16k^2, 40k$
16. $9m, 15n$
17. $7pq, 8q$
18. $18p, 45$

Factor each expression.

19. $5b + 15$
20. $7t + 49$
21. $6w + 18$
22. $100 + 50x$
23. $7x + 7$
24. $12n + 60$
25. $24 + 8g$
26. $50 + 5f$
27. $3n + 24$
28. $9\ell + 63$
29. $6u + 36$
30. $70 - 7c$
31. $42 - 21x$
32. $12y + 16$
33. $6p - 12$
34. $9r - 81$
35. $6 + 8q$
36. $21x + 33$

4-5 Skills Practice

Simplifying Algebraic Fractions

Write each fraction in simplest form. If the fraction is already in simplest form, write *simplified*.

1. $\dfrac{10}{70}$

2. $\dfrac{12}{18}$

3. $\dfrac{30}{45}$

4. $\dfrac{8}{24}$

5. $\dfrac{4}{6}$

6. $\dfrac{56}{63}$

7. $\dfrac{18}{24}$

8. $\dfrac{7}{49}$

9. $\dfrac{13}{39}$

10. $\dfrac{21}{36}$

11. $\dfrac{32}{40}$

12. $\dfrac{4}{36}$

13. $\dfrac{44}{55}$

14. $\dfrac{4}{14}$

15. $\dfrac{36}{48}$

16. $\dfrac{81}{90}$

17. $\dfrac{5}{25}$

18. $\dfrac{56}{74}$

19. $\dfrac{22}{42}$

20. $\dfrac{7}{18}$

21. $\dfrac{d^3}{d^4}$

22. $\dfrac{y}{y^3}$

23. $\dfrac{q^3}{q}$

24. $\dfrac{s^4}{s^2}$

25. $\dfrac{x^2}{y}$

26. $\dfrac{9a}{12a}$

27. $\dfrac{8t}{16t}$

28. $\dfrac{14g}{24g}$

29. $\dfrac{35j}{40}$

30. $\dfrac{100p}{200p^2}$

31. $\dfrac{75n}{100n^3}$

32. $\dfrac{6k^5}{21k^2}$

33. $\dfrac{3a}{4b}$

34. $\dfrac{16b}{24d}$

35. $\dfrac{8a}{24a}$

36. $\dfrac{5t^3}{35t^2}$

4-6 Skills Practice

Multiplying and Dividing Monomials

Find each product or quotient. Express your answer using exponents.

1. $2^3 \cdot 2^5$

2. $10^2 \cdot 10^7$

3. $1^4 \cdot 1$

4. $6^3 \cdot 6^3$

5. $(-3)^2(-3)^3$

6. $(-9)^2(-9)^2$

7. $a^2 \cdot a^3$

8. $n^8 \cdot n^3$

9. $(p^4)(p^4)$

10. $(z^6)(z^7)$

11. $(6b^3)(3b^4)$

12. $(-v)^3(-v)^7$

13. $11a^2 \cdot 3a^6$

14. $10t^2 \cdot 4t^{10}$

15. $(8c^2)(9c)$

16. $(4f^8)(5f^6)$

17. $\dfrac{5^{10}}{5^2}$

18. $\dfrac{10^6}{10^2}$

19. $\dfrac{7^9}{7^6}$

20. $\dfrac{12^8}{12^3}$

21. $\dfrac{100^9}{100^8}$

22. $\dfrac{(-2)^3}{-2}$

23. $\dfrac{r^8}{r^7}$

24. $\dfrac{z^{10}}{z^8}$

25. $\dfrac{q^8}{q^4}$

26. $\dfrac{g^{12}}{g^8}$

27. $\dfrac{(-y)^7}{(-y)^2}$

28. $\dfrac{(-z)^{12}}{(-z)^5}$

29. the product of two squared and two to the sixth power

30. the quotient of ten to the seventh power and ten cubed

31. the product of y squared and y cubed

32. the quotient of a to the twentieth power and a to the tenth power

4-7 Skills Practice

Negative Exponents

Write each expression using a positive exponent.

1. 3^{-4}

2. 8^{-7}

3. 10^{-4}

4. $(-2)^{-6}$

5. $(-40)^{-3}$

6. $(-17)^{-12}$

7. n^{-10}

8. b^{-8}

9. q^{-5}

10. m^{-4}

11. v^{-11}

12. p^{-2}

Write each fraction as an expression using a negative exponent other than -1.

13. $\dfrac{1}{8^2}$

14. $\dfrac{1}{10^5}$

15. $\dfrac{1}{2^3}$

16. $\dfrac{1}{6^7}$

17. $\dfrac{1}{17^4}$

18. $\dfrac{1}{21^2}$

19. $\dfrac{1}{3^7}$

20. $\dfrac{1}{9^2}$

21. $\dfrac{1}{3^2}$

22. $\dfrac{1}{121}$

23. $\dfrac{1}{25}$

24. $\dfrac{1}{36}$

Evaluate each expression if $x = 1$, $y = 2$, and $z = -3$.

25. y^{-z}

26. z^{-2}

27. x^{-8}

28. y^{-5}

29. z^{-3}

30. y^{-1}

31. z^{-4}

32. 5^z

33. x^{-99}

34. 1^z

35. 4^z

36. y^z

4-8 **Skills Practice**

Scientific Notation

Express each number in standard form.

1. 1.5×10^3

2. 4.01×10^4

3. 6.78×10^2

4. 5.925×10^6

5. 7.0×10^8

6. 9.99×10^7

7. 3.0005×10^5

8. 2.54×10^5

9. 1.75×10^4

10. 1.2×10^{-6}

11. 7.0×10^{-1}

12. 6.3×10^{-3}

13. 5.83×10^{-2}

14. 8.075×10^{-4}

15. 1.1×10^{-5}

16. 7.3458×10^7

Express each number in scientific notation.

17. 1,000,000

18. 17,400

19. 500

20. 803,000

21. 0.00027

22. 5300

23. 18

24. 0.125

25. 17,000,000,000

26. 0.01

27. 21,800

28. 2,450,000

29. 0.0054

30. 0.000099

31. 8,888,800

32. 0.00912

Choose the greater number in each pair.

33. $8.8 \times 10^3, 9.1 \times 10^{-4}$

34. $5.01 \times 10^2, 5.02 \times 10^{-1}$

35. $6.4 \times 10^3, 900$

36. $1.9 \times 10^{-2}, 0.02$

37. $2.2 \times 10^{-3}, 2.1 \times 10^2$

38. $8.4 \times 10^2, 839$

5-1 Skills Practice

Writing Fractions as Decimals

Write each fraction or mixed number as a decimal. Use a bar to show a repeating decimal.

1. $\dfrac{3}{4}$

2. $\dfrac{2}{5}$

3. $\dfrac{5}{10}$

4. $\dfrac{5}{5}$

5. $\dfrac{13}{100}$

6. $\dfrac{4}{5}$

7. $\dfrac{7}{10}$

8. $\dfrac{7}{8}$

9. $\dfrac{1}{4}$

10. $\dfrac{7}{50}$

11. $-\dfrac{3}{10}$

12. $\dfrac{1}{11}$

13. $1\dfrac{1}{8}$

14. $\dfrac{1}{12}$

15. $\dfrac{7}{30}$

16. $\dfrac{1}{15}$

17. $4\dfrac{7}{11}$

18. $-10\dfrac{5}{9}$

19. $-2\dfrac{3}{5}$

20. $6\dfrac{1}{6}$

Replace each ● with <, >, or = to make a true sentence.

21. $\dfrac{1}{8}$ ● 0.12

22. $\dfrac{2}{3}$ ● 0.7

23. $-2\dfrac{3}{10}$ ● -2.3

24. 0.395 ● $\dfrac{2}{5}$

25. 0.1 ● $\dfrac{1}{11}$

26. $0.\overline{16}$ ● $\dfrac{1}{6}$

27. $\dfrac{3}{5}$ ● $\dfrac{3}{4}$

28. $-3\dfrac{1}{4}$ ● -3.25

29. Order $\dfrac{9}{11}$, 0.99, and $\dfrac{9}{10}$ from least to greatest.

30. Order 0.5, $\dfrac{4}{9}$, and $\dfrac{2}{5}$ from least to greatest.

5-2 Skills Practice

Rational Numbers

Write each number as a fraction.

1. 13

2. $1\frac{1}{4}$

3. 57

4. -25

5. $-3\frac{4}{5}$

6. $6\frac{5}{8}$

7. -1

8. $2\frac{2}{9}$

Write each decimal as a fraction or mixed number in simplest form.

9. 0.6

10. 0.25

11. $0.\overline{4}$

12. $-1.\overline{1}$

13. 0.11

14. 2.8

15. 7.03

16. -2.12

17. $3.\overline{2}$

18. 1.125

19. 8.65

20. 16.7

21. 0.16

22. 4.06

23. $-5.\overline{8}$

24. $0.\overline{24}$

25. Write 85 hundredths as a fraction in simplest form.

26. Write 9 and 250 thousandths as a mixed number in simplest form.

Identify all sets to which each number belongs (N = natural numbers, W = whole numbers, I = integers, Q = rational numbers).

27. 16

28. -2.54

29. $\frac{9}{3}$

30. $0.\overline{95}$

31. -4

32. 2.2020020002...

5-3 **Skills Practice**

Multiplying Rational Numbers

Find each product. Write in simplest form.

1. $\dfrac{1}{3} \cdot \left(-\dfrac{1}{4}\right)$

2. $-\dfrac{2}{5} \cdot \dfrac{6}{7}$

3. $\dfrac{2}{7} \cdot \dfrac{3}{11}$

4. $\dfrac{3}{13} \cdot \dfrac{2}{5}$

5. $\dfrac{2}{9} \cdot \dfrac{3}{5}$

6. $\dfrac{3}{11} \cdot \dfrac{5}{9}$

7. $-\dfrac{1}{4} \cdot \dfrac{4}{9}$

8. $\dfrac{3}{5} \cdot \dfrac{15}{18}$

9. $\dfrac{5}{16} \cdot 4$

10. $5\dfrac{1}{2} \cdot \dfrac{2}{11}$

11. $-12\dfrac{2}{3} \cdot 7\dfrac{1}{2}$

12. $-\dfrac{5}{36} \cdot \left(-\dfrac{9}{25}\right)$

13. $8\dfrac{4}{5} \cdot 2\dfrac{5}{10}$

14. $3\dfrac{1}{3} \cdot 9\dfrac{3}{4}$

15. $\dfrac{3}{a} \cdot \dfrac{b}{5}$

16. $\dfrac{2x}{5} \cdot \dfrac{3}{x}$

17. $\dfrac{9m}{n} \cdot \dfrac{2n}{3}$

18. $\dfrac{3s}{t^2} \cdot \dfrac{t}{9s^2}$

19. $\dfrac{ab}{c} \cdot \dfrac{c^2}{b}$

20. $\dfrac{x}{y} \cdot \dfrac{2x}{4}$

21. $\dfrac{r^2s}{t} \cdot \dfrac{3t^2}{rs^3}$

22. $\dfrac{ab^3}{9} \cdot \dfrac{b}{a^2}$

23. $\dfrac{m^4n^2p}{4} \cdot \dfrac{8p^2}{m^4n}$

24. $\dfrac{3xy}{5} \cdot \dfrac{15xyz^2}{y^2}$

MEASUREMENT   Complete.

25. ____?____ quarts = $6\dfrac{1}{4}$ gallons

26. $9\dfrac{3}{5}$ minutes = ____?____ seconds

27. ____?____ days = $17\dfrac{1}{2}$ weeks

5-4 Skills Practice

Dividing Rational Numbers

Find the multiplicative inverse of each number.

1. $\dfrac{7}{12}$

2. $-\dfrac{3}{10}$

3. $\dfrac{1}{8}$

4. -64

5. $8\dfrac{1}{3}$

6. $-10\dfrac{2}{3}$

7. $-6\dfrac{5}{6}$

8. $1\dfrac{1}{8}$

Find each quotient. Write in simplest form.

9. $\dfrac{1}{3} \div \dfrac{7}{18}$

10. $-\dfrac{2}{5} \div \dfrac{4}{25}$

11. $-5 \div \dfrac{1}{7}$

12. $\dfrac{2}{3} \div \dfrac{2}{3}$

13. $\dfrac{4}{5} \div \left(-\dfrac{1}{15}\right)$

14. $\dfrac{19}{20} \div \dfrac{4}{5}$

15. $3 \div \dfrac{1}{4}$

16. $-15 \div \dfrac{1}{2}$

17. $\dfrac{4}{9} \div \dfrac{5}{12}$

18. $\dfrac{7}{10} \div \left(-\dfrac{4}{5}\right)$

19. $\dfrac{7}{12} \div \left(-1\dfrac{1}{6}\right)$

20. $1\dfrac{5}{8} \div \dfrac{5}{8}$

21. $12\dfrac{3}{5} \div 2\dfrac{7}{10}$

22. $-\dfrac{3}{11} \div \dfrac{6}{22}$

23. $\dfrac{1}{8} \div \dfrac{15}{16}$

24. $-12\dfrac{4}{5} \div \left(-1\dfrac{1}{15}\right)$

25. $1\dfrac{12}{13} \div \dfrac{25}{26}$

26. $-7\dfrac{1}{3} \div 2\dfrac{1}{5}$

27. $\dfrac{x}{6} \div \dfrac{x}{30}$

28. $\dfrac{12}{5x} \div \dfrac{6}{2x}$

29. $\dfrac{m}{16} \div \dfrac{mp}{7}$

30. $\dfrac{3r}{s} \div \dfrac{4rs}{s^2}$

31. $\dfrac{a}{b} \div \dfrac{5}{b}$

32. $\dfrac{2a}{b} \div \dfrac{3a^2}{b^2}$

33. $\dfrac{3}{5c} \div \dfrac{1}{10c}$

34. $\dfrac{pq}{6} \div \dfrac{q}{8}$

35. $\dfrac{x^2}{7} \div \dfrac{2x}{21}$

36. $\dfrac{gh}{6} \div \dfrac{36}{h}$

37. $\dfrac{3n}{2m} \div \dfrac{5n}{5m}$

38. $\dfrac{4b}{c} \div \dfrac{5bc}{c}$

Glencoe Pre-Algebra

5-5 Skills Practice

Adding and Subtracting Like Fractions

Find each sum or difference. Write in simplest form.

1. $\dfrac{4}{15} + \dfrac{6}{15}$

2. $\dfrac{7}{12} + \dfrac{11}{12}$

3. $\dfrac{7}{10} + \dfrac{9}{10}$

4. $\dfrac{20}{21} - \dfrac{2}{21}$

5. $\dfrac{11}{12} - \dfrac{5}{12}$

6. $\dfrac{5}{8} + \dfrac{7}{8}$

7. $\dfrac{10}{11} + \dfrac{9}{11}$

8. $\dfrac{17}{30} - \dfrac{7}{30}$

9. $\dfrac{5}{6} + \dfrac{5}{6}$

10. $4\dfrac{4}{5} + 3\dfrac{2}{5}$

11. $20\dfrac{1}{25} + 1\dfrac{4}{25}$

12. $5\dfrac{11}{15} + 3\dfrac{14}{15}$

13. $26\dfrac{7}{12} + 11\dfrac{11}{12}$

14. $20\dfrac{3}{4} - 3\dfrac{1}{4}$

15. $25\dfrac{4}{5} - 3\dfrac{2}{5}$

16. $\dfrac{10}{15} - \dfrac{13}{15}$

17. $\dfrac{a}{6} + \dfrac{4a}{6}$

18. $\dfrac{7c}{16} + \dfrac{7c}{16}$

19. $\dfrac{25}{x} - \dfrac{17}{x}, x \neq 0$

20. $1\dfrac{1}{2y} - 2\dfrac{1}{2y}$

21. $\dfrac{7x}{9} - \dfrac{7x}{9}$

22. $\dfrac{3m}{5} + \dfrac{8m}{5}$

Evaluate each expression if $x = \dfrac{5}{8}$, $y = 1\dfrac{3}{8}$, and $z = \dfrac{1}{8}$.

23. $x + y$

24. $y - x$

25. $x - z$

26. $x + y + z$

5-6 Skills Practice

Least Common Multiple

Find the least common multiple (LCM) of each pair of numbers or monomials.

1. 9, 12

2. 8, 10

3. 7, 21

4. 6, 10

5. 14, 35

6. 18, 24

7. 30, 12

8. 36, 42

9. $5a, 3a^2$

10. $8st, 5t$

11. rs, s^2

12. $2b^2, 3ab$

Find the least common denominator (LCD) of each pair of fractions.

13. $\dfrac{2}{3}, \dfrac{3}{4}$

14. $\dfrac{7}{20}, \dfrac{8}{22}$

15. $\dfrac{3}{10}, \dfrac{7}{12}$

16. $\dfrac{4}{17}, \dfrac{23}{51}$

17. $\dfrac{7}{9}, \dfrac{5}{6}$

18. $\dfrac{73}{100}, \dfrac{73}{80}$

19. $\dfrac{7}{48}, \dfrac{25}{36}$

20. $\dfrac{7}{3p}, \dfrac{6}{2p}$

21. $\dfrac{8}{d^2e}, \dfrac{7}{9d^3}$

22. $\dfrac{9}{5ab}, \dfrac{7}{10c^2}$

Replace each ● with <, >, or = to make a true sentence.

23. $-\dfrac{5}{6}$ ● $\dfrac{5}{15}$

24. $\dfrac{3}{4}$ ● $\dfrac{9}{10}$

25. $\dfrac{3}{5}$ ● $\dfrac{4}{7}$

26. $\dfrac{1}{8}$ ● $\dfrac{4}{32}$

27. $\dfrac{3}{9}$ ● $\dfrac{5}{27}$

28. $\dfrac{11}{16}$ ● $\dfrac{22}{24}$

29. $\dfrac{2}{9}$ ● $\dfrac{3}{10}$

30. $\dfrac{1}{6}$ ● $\dfrac{4}{25}$

5-7 Skills Practice

Adding and Subtracting Unlike Fractions

Find each sum or difference. Write in simplest form.

1. $\dfrac{4}{7} + \dfrac{1}{3}$

2. $\dfrac{2}{5} + \dfrac{3}{4}$

3. $\dfrac{1}{2} + \left(-\dfrac{3}{10}\right)$

4. $-\dfrac{5}{6} + \dfrac{7}{9}$

5. $\dfrac{5}{12} + \dfrac{23}{24}$

6. $\dfrac{10}{11} - \dfrac{1}{2}$

7. $\dfrac{4}{5} - \left(-\dfrac{1}{3}\right)$

8. $\dfrac{5}{6} - \dfrac{1}{12}$

9. $\dfrac{19}{20} + \dfrac{1}{4}$

10. $-\dfrac{9}{10} - \dfrac{1}{3}$

11. $\dfrac{13}{15} - \dfrac{2}{3}$

12. $\dfrac{7}{10} + \dfrac{1}{5}$

13. $-\dfrac{3}{8} + \dfrac{1}{6}$

14. $\dfrac{33}{100} - \dfrac{1}{10}$

15. $\dfrac{11}{12} - \left(-\dfrac{7}{8}\right)$

16. $\dfrac{4}{5} - \dfrac{1}{8}$

17. $5\dfrac{2}{3} + 2\dfrac{1}{6}$

18. $1\dfrac{7}{8} + 3\dfrac{1}{3}$

19. $3\dfrac{2}{3} - \dfrac{1}{9}$

20. $23\dfrac{3}{4} - 12\dfrac{5}{16}$

21. $-7\dfrac{1}{2} + \dfrac{3}{4}$

22. $2\dfrac{8}{12} + 1\dfrac{1}{4}$

23. $-12\dfrac{1}{2} - 17\dfrac{1}{2}$

24. $12\dfrac{1}{3} - \dfrac{3}{5}$

25. $11\dfrac{15}{16} - 7\dfrac{1}{2}$

26. $8\dfrac{5}{9} + 1\dfrac{1}{6}$

27. $-7\dfrac{1}{2} + 3\dfrac{1}{7}$

28. $60\dfrac{1}{2} + \left(-37\dfrac{1}{6}\right)$

29. $8\dfrac{2}{3} - 3\dfrac{1}{3}$

30. $-21\dfrac{7}{16} + 13\dfrac{1}{4}$

Glencoe Pre-Algebra

5-8 Skills Practice

Measures of Central Tendency

Find the mean, median, and mode for each set of data. If necessary, round to the nearest tenth.

1. 6, 3, 3, 12, 13, 15, 7

2. 1, 1, 0, 2, 1, 1, 0, 0, 1

3. 202, 195, 219, 220

4. 2.5, 4.0, 8.7, 3.3, 3.3, 5.2

5. 21, 23, 39, 44, 27, 25, 28, 30

6. 87, 85, 87, 87, 87

Find the mean, median, and mode for each set of data. If necessary, round to the nearest tenth.

7.

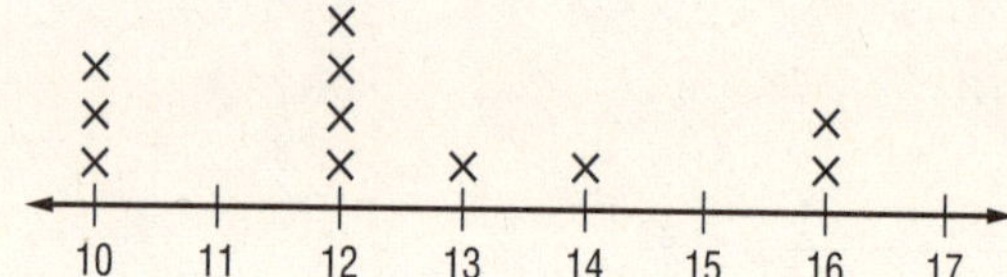

8.

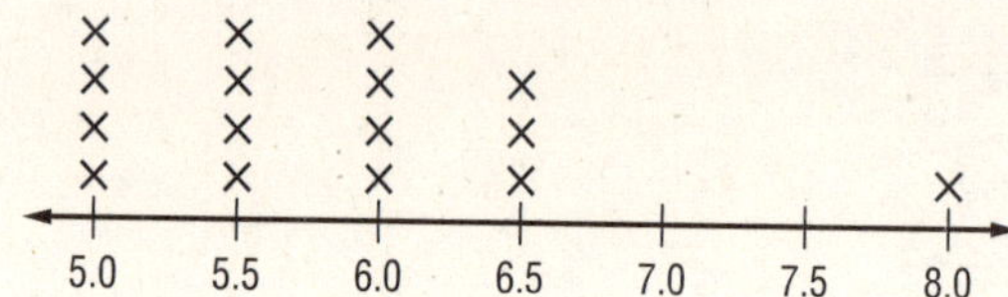

9. TEMPERATURE The average daily temperature by month for one year in Denver, Colorado, is given below. Find the mean, median, and mode for temperature.

Month	Jan	Feb	Mar	Apr	May	June	July	Aug	Sept	Oct	Nov	Dec
Temp. (°F)	43°	47°	51°	61°	71°	82°	88°	86°	78°	67°	52°	46°

Source: *The Universal Almanac*

5-9 Skills Practice

Solving Equations with Rational Numbers

Solve each equation. Check your solution.

1. $b + 2.7 = 3.5$

2. $c - 9.8 = 3.6$

3. $\dfrac{1}{6} + d = \dfrac{3}{4}$

4. $x - \dfrac{8}{9} = 1$

5. $-3.2 + y = 8.4$

6. $3.5x = 14$

7. $\dfrac{2}{7} + p = -\dfrac{3}{4}$

8. $2x = -9.0$

9. $\dfrac{x}{6} = 10.2$

10. $g - 8.95 = 11.45$

11. $h + \dfrac{6}{11} = \dfrac{3}{22}$

12. $a + 6.11 = 8.45$

13. $\dfrac{2}{3}m = \dfrac{20}{27}$

14. $w - 3\dfrac{2}{3} = 4\dfrac{5}{6}$

15. $0.22f = 13.2$

16. $\dfrac{x}{0.5} = 18$

17. $3.14r = 7.85$

18. $\dfrac{7}{8}s = \dfrac{4}{5}$

19. $-3.9 + r = -8.5$

20. $9.8a = 20.58$

21. $\dfrac{9}{100} + h = \dfrac{1}{10}$

22. $\dfrac{m}{8} = \dfrac{15}{24}$

23. $\dfrac{4}{5}t = 8$

24. $10 = \dfrac{3}{4}s$

25. $-0.5v = -9$

26. $3x = 1\dfrac{1}{2}$

27. $v - 1\dfrac{1}{3} = 1\dfrac{4}{5}$

28. $3\dfrac{7}{8} + q = 1\dfrac{1}{2}$

29. $\dfrac{5}{11}d = 35$

30. $4\dfrac{2}{3} = \dfrac{1}{3}y$

Glencoe Pre-Algebra

5-10 Skills Practice

Arithmetic and Geometric Sequences

State whether each sequence is *arithmetic*, *geometric*, or *neither*. If it is arithmetic or geometric, state the common difference or common ratio and write the next three terms of the sequence.

1. 5, 9, 13, 17, 21, . . .

2. 15, 10, 5, 0, −5, . . .

3. 1, 3, 9, 27, 81, . . .

4. 1, 0.3, 0.09, 0.027, 0.0081, . . .

5. $\frac{1}{2}$, 1, $\frac{1}{2}$, 1, $\frac{1}{2}$, . . .

6. 0.4, 1.6, 6.4, 25.6, 102.4, . . .

7. 2, 1, $\frac{1}{2}$, $\frac{1}{4}$, $\frac{1}{8}$, . . .

8. 30, 21, 12, 3, −6, . . .

9. $\frac{1}{16}$, $\frac{1}{8}$, $\frac{1}{4}$, $\frac{1}{2}$, 1, . . .

10. 3, −4, −11, −18, −25, . . .

11. 1, 2, 4, 7, 11, . . .

12. 0, 9, 18, 27, 36, . . .

13. 100, 10, 1, 0.1, 0.01, . . .

14. 1.0, 1.2, 1.4, 1.6, 1.8, . . .

6-1 Skills Practice

Ratios and Rates

Express each ratio as a fraction in simplest form.

1. 8 pencils out of 12 pens

2. 42 textbooks to 28 students

3. 27 rooms to 48 windows

4. 15 angel fish to 75 fish

5. 75 cats to 100 dogs

6. 6 aces out of 24 serves

7. 3 gallons to 15 quarts

8. 30 feet to 11 yards

Express each ratio as a unit rate. Round to the nearest tenth, if necessary.

9. $9 for 6 cans of soup

10. $39 for a case of 75 bananas

11. 108 miles in 6 days

12. 51 meters in 8 seconds

13. 21 new pairs of sneakers in 7 years

14. 52 feet for 8 costumes

15. 40 sneezes in 20 minutes

16. $2702 from 28 people

Convert each rate using dimensional analysis.

17. 12 m/min = __?__ cm/s

18. 8 qt/min = __?__ gal/h

19. 44 yd/s = __?__ mi/h

20. 10 c/min = __?__ qt/h

21. 32 ft/h = __?__ yd/day

22. 56 mi/h = __?__ ft/min

23. 40 cm/s = __?__ m/min

24. 180 in./min = __?__ yd/h

6-2 Skills Practice

Using Proportions

Determine whether each pair of ratios forms a proportion.

1. $\dfrac{1}{5}, \dfrac{4}{20}$

2. $\dfrac{3}{8}, \dfrac{12}{32}$

3. $\dfrac{4}{5}, \dfrac{9}{10}$

4. $\dfrac{12}{20}, \dfrac{18}{30}$

5. $\dfrac{3}{4}, \dfrac{27}{36}$

6. $\dfrac{10}{18}, \dfrac{2}{9}$

7. $\dfrac{4}{9}, \dfrac{2}{3}$

8. $\dfrac{15}{18}, \dfrac{10}{12}$

9. $\dfrac{15}{24}, \dfrac{3}{8}$

10. $\dfrac{36}{72}, \dfrac{50}{100}$

11. $\dfrac{10}{8.4}, \dfrac{5}{4.2}$

12. $\dfrac{12}{4.8}, \dfrac{9}{3.2}$

ALGEBRA Solve each proportion.

13. $\dfrac{8}{4} = \dfrac{t}{8}$

14. $\dfrac{n}{9} = \dfrac{4}{18}$

15. $\dfrac{3}{v} = \dfrac{12}{32}$

16. $\dfrac{25}{60} = \dfrac{s}{12}$

17. $\dfrac{21}{28} = \dfrac{3}{w}$

18. $\dfrac{c}{12} = \dfrac{5}{6}$

19. $\dfrac{4}{r} = \dfrac{5}{20}$

20. $\dfrac{12}{18} = \dfrac{m}{81}$

21. $\dfrac{2}{9} = \dfrac{6}{k}$

22. $\dfrac{h}{35} = \dfrac{3}{7}$

23. $\dfrac{3}{16} = \dfrac{u}{40}$

24. $\dfrac{6}{a} = \dfrac{1}{3}$

25. $\dfrac{e}{9.5} = \dfrac{6.4}{7.6}$

26. $\dfrac{2.7}{3.0} = \dfrac{3.6}{x}$

27. $\dfrac{1.68}{w} = \dfrac{7}{12}$

Glencoe Pre-Algebra

6-3 **Skills Practice**

Scale Drawings and Models

On a set of architectural drawings for a new school building, the scale is $\frac{1}{4}$ inch = 2 feet. Find the missing lengths of the rooms.

	Room	Drawing Length	Actual Length
1.	Lobby		16 feet
2.	Principal's Office	1.25 inches	
3.	Library		20 feet
4.	School Room	3 inches	
5.	Science Lab	1.5 inches	
6.	Cafeteria		48 feet
7.	Music Room	4 inches	
8.	Gymnasium	13 inches	
9.	Auditorium		56 feet
10.	Teachers' Lounge	1.75 inches	

11. Refer to Exercises 1–10. What is the scale factor?

12. What is the scale factor if the scale is 10 inches = 1 foot?

13. STRUCTURES A barn is 40 feet wide by 100 feet long. Make a scale drawing of the barn that has a scale of $\frac{1}{2}$ inch = 10 feet.

14. MAPS On a map, the key indicates that 1 centimeter equals 3.5 meters. A road is shown on this map that runs for 30 centimeters. How long is this road?

6-4 Skills Practice

Fractions, Decimals, and Percents

Express each percent as a fraction or mixed number in simplest form and as a decimal.

1. 55%   2. 2%   3. $5\frac{1}{2}\%$ 4. 30%

5. 300%   6. 12%   7. 50%   8. 90%

9. 85%   10. 28.2%   11. 0.25%   12. 0.2%

13. 7.5%   14. 6%   15. 10%   16. 275%

Express each decimal or fraction as a percent. Round to the nearest tenth percent, if necessary.

17. 0.65   18. 0.772   19. 0.6   20. 3.45

21. 0.47   22. 0.01   23. 22.6   24. 0.79

25. 0.28   26. 0.355   27. 0.0015   28. 44

29. $\frac{11}{20}$ 30. $\frac{1}{4}$ 31. $\frac{5}{8}$ 32. $\frac{7}{5}$

33. $\frac{23}{4}$ 34. $\frac{4}{5}$ 35. $\frac{3}{25}$ 36. $\frac{7}{3}$

37. $2\frac{3}{10}$ 38. $\frac{1}{6}$ 39. $\frac{300}{630}$ 40. $\frac{9}{10}$

6-5 Skills Practice

Using the Percent Proportion

Use the percent proportion to solve each problem. Round to the nearest tenth.

1. 64 is what percent of 200?

2. What percent of 12 is 9?

3. 2 is what percent of 80?

4. What percent of 42 is 32?

5. 10 is what percent of 60?

6. What percent of 30 is 6?

7. 15 is what percent of 24?

8. What percent of 36 is 9?

9. 28 is what percent of 42?

10. What percent of 72 is 21?

11. 8 is 40% of what number?

12. 16 is 5% of what number?

13. 25 is 80% of what number?

14. 0.84 is 28% of what number?

15. 71 is 10% of what number?

16. 52 is 97% of what number?

17. 39 is 17% of what number?

18. 12 is 4% of what number?

19. 48.5 is 7% of what number?

20. What is 10.6% of 11?

21. What is 15% of 98.4?

22. What is 0.5% of 75?

23. What is 4% of 512.5?

24. What is 50% of 1?

25. What is 25% of 12?

26. What is 12% of 25?

27. What is 90% of 50?

28. What is 50% of 90?

6-6 Skills Practice

Finding Percents Mentally

Find the percent of each number mentally.

1. 10% of 582

2. 50% of 86

3. 40% of 1500

4. 20% of 75

5. 15% of 20

6. 80% of 45

7. 30% of 120

8. 75% of 44

9. 5% of 40

10. $33\frac{1}{3}$% of 99

11. 60% of 450

12. $37\frac{1}{2}$% of 56

13. 25% of 480

14. 300% of 5

15. 150% of 82

16. $66\frac{2}{3}$% of 210

17. 125% of 800

18. 175% of 400

Estimate.

19. 28% of 19

20. 55% of 32

21. 87% of 158

22. 35% of 544

23. 42% of 495

24. 19% of 319

25. 65% of 73

26. 8% of 224

27. 83% of 9

28. 17% of 331

29. 78% of 14

30. 12% of 879

31. $\frac{1}{3}$% of 941

32. $\frac{1}{2}$% of 376

33. $\frac{1}{5}$% of 2052

34. 164% of 318

35. 247% of 192

36. 508% of 1073

6-7 Skills Practice

Using Percent Equations

Solve each problem using the percent equation.

1. What is 5% of 80?

2. What is 10% of 100?

3. What is 58% of 35?

4. What is 32% of 150?

5. What is 91% of 3800?

6. Find 25% of 68.

7. Find 80% of 75.

8. Find 75% of 80.

9. Find 1.5% of 8400.

10. Find 33.5% of 22.

11. 23 is what percent of 115?

12. 27 is what percent of 75?

13. 80 is what percent of 160?

14. 85 is what percent of 500?

15. 48 is what percent of 30?

16. 321.3 is what percent of 918?

17. 0.6 is what percent of 2?

18. 126 is what percent of 140?

19. 21 is what percent of 1050?

20. 78 is what percent of 40?

21. 29 is 50% of what number?

22. 9 is 45% of what number?

23. 16 is 4% of what number?

24. 336 is 48% of what number?

25. 52 is 25% of what number?

26. 99 is 90% of what number?

27. 343 is 70% of what number?

28. 57 is 1% of what number?

29. 193.6 is 32% of what number?

30. 87.1 is 67% of what number?

6-8 Skills Practice

Percent of Change

State whether each change is a *percent of increase* or a *percent of decrease*. Then find the percent of change. Round to the nearest tenth, if necessary.

1. from 12 m to 18 m

2. from 27 days to 30 days

3. from $48.50 to $38.80

4. from 25 lb to 12 lb

5. from 10 mm to 3 mm

6. from $875 to $1000

7. from $18.10 to $22.50

8. from 32 people to 3040 people

9. from 28 stray cats to 5 stray cats

10. from 12 words to 90 words

11. from 47 mph to 35 mph

12. from 8 computers to 15 computers

13. from 34 workers to 28 workers

14. from 8056 snowflakes to 6381 snowflakes

15. from 201 sales to 148 sales

16. from 153 balls to 380 balls

17. from 5 miles to 8 miles

18. from 850 singers to 715 singers

19. from 9 horses to 11 horses

20. from 900 CDs to 1100 CDs

21. from 14 cheerleaders to 12 cheerleaders

22. from 140 members to 120 members

23. from $200 to $210

24. from $210 to $200

25. from 300 s to 8 s

26. from 8 s to 300 s

6-9 Skills Practice

Probability and Predictions

A spinner like the one shown is used in a game.
Determine the probability of each outcome if
the spinner is equally likely to land on each
section. Express each probability as a fraction
and as a percent.

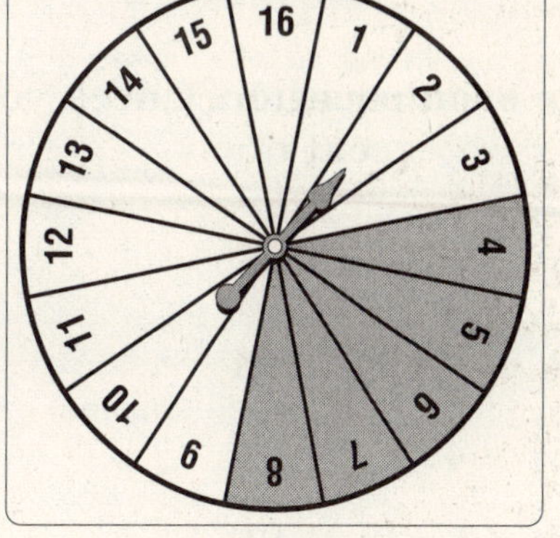

1. $P(10)$ **2.** $P(\text{odd})$

3. $P(\text{greater than 7})$ **4.** $P(\text{prime})$ **5.** $P(1 \text{ or } 2)$

6. $P(\text{less than 5})$ **7.** $P(\text{Shaded})$ **8.** $P(\text{Not shaded})$

There are 4 red marbles, 1 blue marble, 9 green marbles, and 6 yellow marble in
a bag. Suppose one marble is selected at random. Find the probability of each
outcome. Express each probability as a fraction and as a percent.

9. $P(\text{red})$ **10.** $P(\text{blue})$ **11.** $P(\text{yellow})$

12. $P(\text{red or blue})$ **13.** $P(\text{white})$ **14.** $P(\text{red, blue, or green})$

Suppose two 1–6 number cubes are rolled. Find the probability of each outcome.
Express each probability as a fraction and as a percent. (*Hint:* Make a table to
show the sample space as in Example 2.) Round to the nearest tenth, if necessary.

15. $P(3 \text{ or } 5)$ **16.** $P(\text{both even})$ **17.** $P(\text{odd product})$

18. $P(\text{sum more than 10})$ **19.** $P(\text{both the same})$ **20.** $P(\text{product is a square})$

7-1 Skills Practice

Solving Equations with Variables on Each Side

Solve each equation. Check your solution.

1. $3x + 2 = 5x$

2. $n - 12 = 3n$

3. $2 - 3b = 7b + 12$

4. $4d - 11 = 2d + 7$

5. $2f + 3 = 11f - 24$

6. $8y + 11 = 2y + 29$

7. $5a = 45 + 2a$

8. $17 - 3c = 4c + 3$

9. $2a - 3 = 9a - 10$

10. $5b = 21 + 4b$

11. $9y - 27 = -2y + 6$

12. $2n - 5 = 7n$

13. $-s + 3 = 5s + 21$

14. $7 - 4c = 3c + 35$

15. $30 - 2n = 4n$

16. $29 + 7d = 5d + 15$

17. $16k - 23 = 6k - 13$

18. $w - 20 = 6w$

19. $33g + 28 = 25g - 12$

20. $6h - 34 = -6h + 14$

21. $3t + 17 = t - 3$

22. $11j = 6j - 15$

23. $c - 2 = 3c + 14$

24. $28x - 7 = 26x + 5$

25. $5m - 6 = 8m + 9$

26. $-4p - 7 = 5p + 11$

27. $-10 + 3f = 5f + 6$

28. $4f + 6 = 8f - 14$

29. $-7n - 16 = 4n + 17$

30. $5d = 9d - 18$

Define a variable and write an equation to find each number. Then solve.

31. Three times a number equals 40 more than five times the number. What is the number?

32. A number equals four less than three times the number. What is the number?

33. Eight times a number equals 24 more than two times the number. What is the number?

7-2 Skills Practice

Solving Equations with Grouping Symbols

Solve each equation. Check your solution.

1. $2(g - 7) = 16$

2. $5(x + 2) = 30$

3. $3(2d + 7) = 39$

4. $4(a - 2) = 3(a + 4)$

5. $3(f + 2) + 9 = 13 + 5f$

6. $2(x - 4) = 3(1 + x)$

7. $2n + 5 = 4(n + 2) - n$

8. $4(x + 3) = x$

9. $2(c - 3) = 76$

10. $7(x - 2) = 5(x + 2)$

11. $2(6x + 1) = 4(x - 5) - 2$

12. $4(2b - 6) + 11 = 8b - 13$

13. $6 + 6(2t - 1) = 3 + 12t$

14. $9t - 21 = 3(t - 7) + 6t$

15. $3(4k + 14) = 10k - 2(k - 7)$

Find the dimensions of each rectangle. The perimeter is given.

16. $P = 380$ m

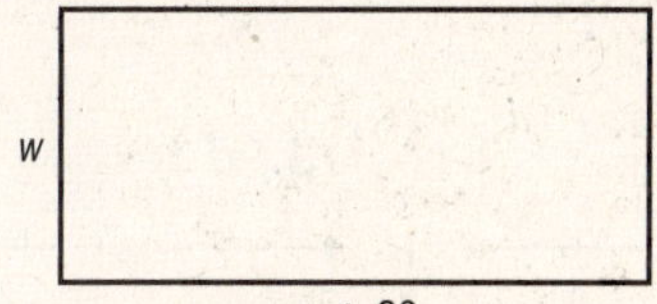

17. $P = 640$ yd

18. $P = 220$ ft

19. $P = 380$ yd

20. $P = 300$ m

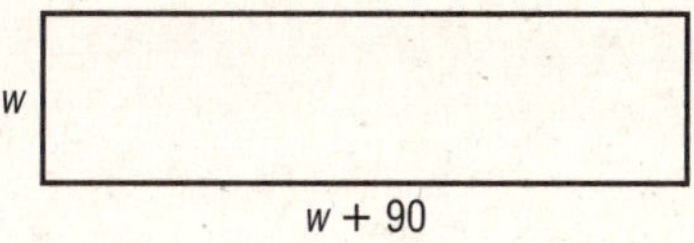

7-3 Skills Practice

Inequalities

Write an inequality for each sentence.

1. More than 100,000 fans attended the opening football game at The Ohio State University.

2. Her earnings at $16 per hour were no more than $96.

3. A savings account decreased by $50 is now less than $740.

4. A number increased by 7 is at least 45.

For the given value, state whether each inequality is true or false.

5. $\dfrac{18}{c} < 9, c = 2$

6. $\dfrac{x}{5} \geq 3, x = 5$

7. $6k \geq 42, k = 7$

8. $10 - x < 3, x = 7$

9. $11 + n < 32, n = 4$

10. $9 + c > 19, c = 10$

Graph each inequality on a number line.

11. $a < 6$
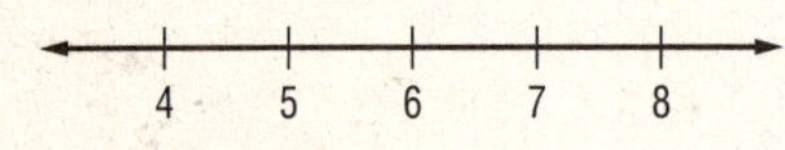

12. $t \geq -2$
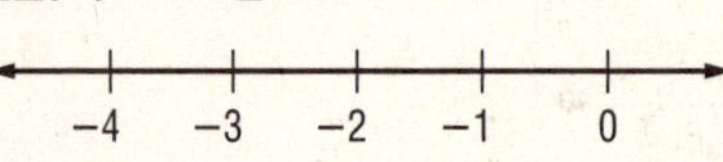

13. $d \leq 3$
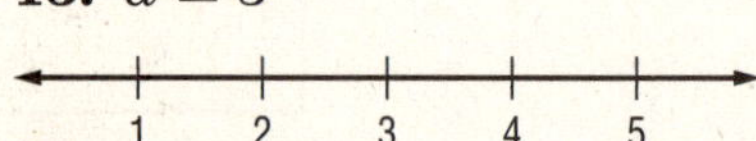

14. $b \geq 10$
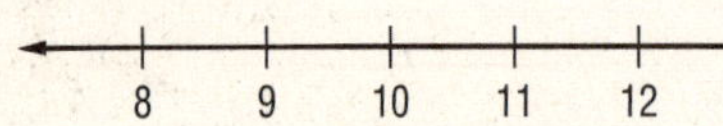

15. $x \geq -7$
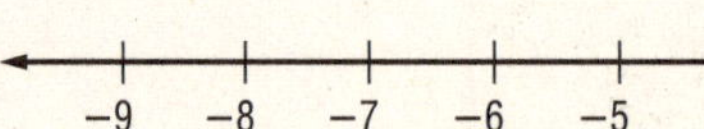

16. $x > 2$
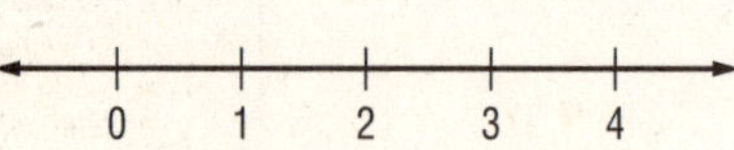

Write the inequality for each graph.

17.
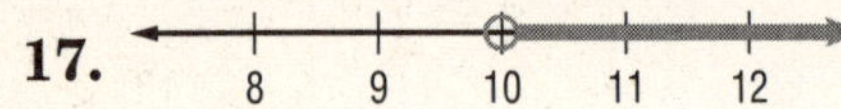

18.
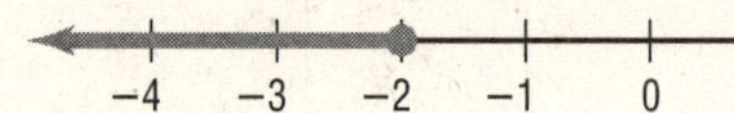

19.
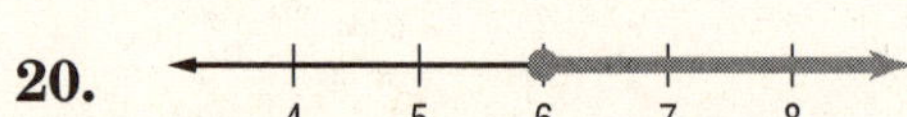

20.

21.
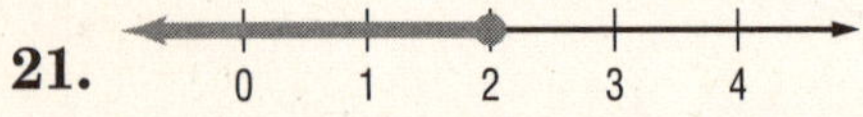

22.

7-4 Skills Practice

Solving Inequalities by Adding or Subtracting

Solve each inequality. Check your solution.

1. $p + 9 > 13$
2. $t + 7 < -4$
3. $-12 \geq 7 + x$
4. $f + (-7) \leq 9$

5. $5 > -3 + y$
6. $r + 7 \leq -3$
7. $b - 15 > 11$
8. $z + (-4) < -8$

9. $j - 4 \leq -10$
10. $-5 > h - 3$
11. $13 > w - (-14)$
12. $g - 7 > -4$

13. $-15 \leq d + (-2)$
14. $2 + c \leq -8$
15. $15 > c + 3$
16. $j + 9 \leq -10$

Solve each inequality. Then graph the solution on a number line.

17. $n + 6 < 7$

18. $-4 + z \geq 3$

19. $p + (-3) > -6$

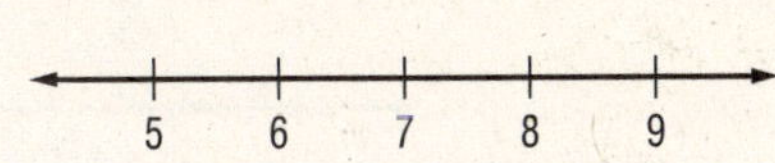
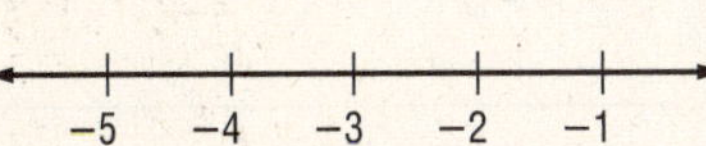

20. $-11 \geq x + 8$

21. $15 \leq m - (-2)$

22. $k + 7 \geq -9$

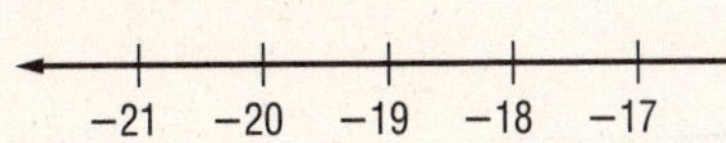
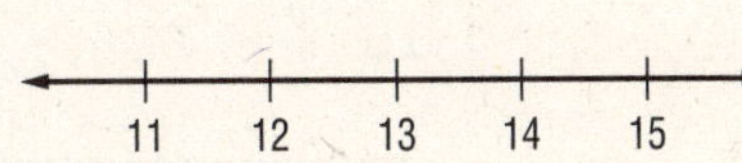
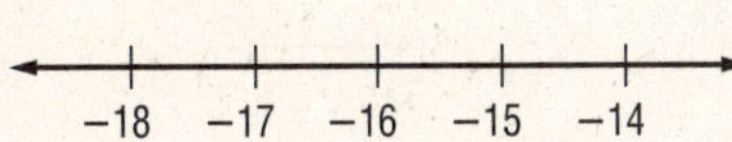

23. $-5 < 2 + a$

24. $t - 7 > 5$

25. $-2 + z \geq 7$

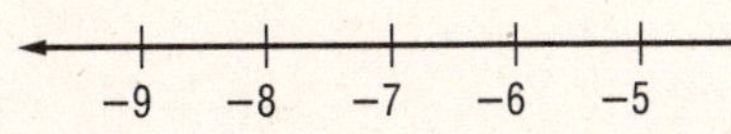
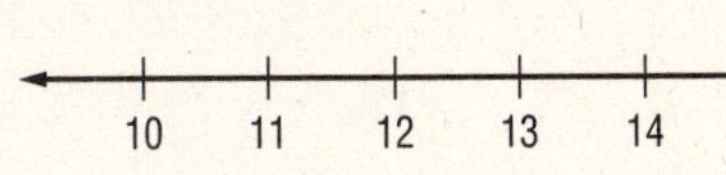
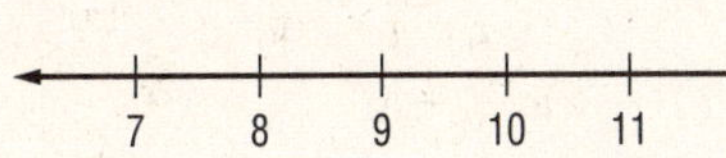

26. $p - (-7) > -8$

27. $-6 \leq m + 2$

28. $-2 < w - 5$

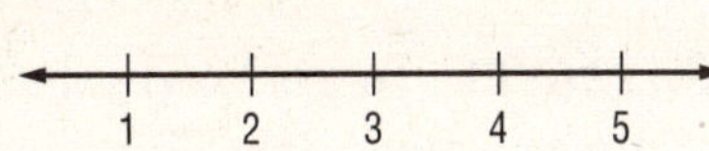

29. **SHOPPING** Chantal would like to buy a new pair of running shoes. Shoes that she likes start at \$85. If she has already saved \$62, what is the least amount she must still save?

7-5 Skills Practice

Solving Inequalities by Multiplying or Dividing

Solve each inequality and check your solution. Then graph the solution on a number line.

1. $-8x > 16$

2. $7y < -35$

3. $12a \geq -24$

4. $-12 \leq 4a$

5. $-6z < -18$

6. $14 > -2k$

7. $5 > \dfrac{x}{-2}$

8. $\dfrac{r}{-3} \leq -4$

9. $-10t \geq 200$

10. $\dfrac{y}{7} < 2$

11. $\dfrac{-1}{2}x \leq -6$

12. $\dfrac{b}{-3} \leq 6$

13. TRAVEL To get to the beach for vacation, Cheng's family must drive at least 660 miles on the first day. They are traveling at a constant speed of 60 miles per hour.

   a. Write an inequality to represent how long the family should drive on the first day.

   b. How many hours should the family drive?

14. EARNINGS Jess receives $180 for every garage he paints over the summer. He wants to save at least $1620 for college.

   a. Write an inequality to represent how many garages Jess should paint over the summer.

   b. How many garages should Jess paint?

7-6 Skills Practice

Solving Multi-Step Inequalities

Solve each inequality and check your solution. Graph the solution on a number line.

1. $3x + 9 < 18$

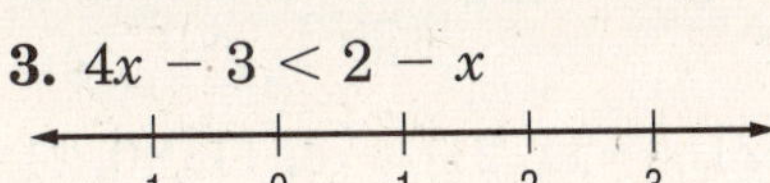

2. $5 + 2c < -9$

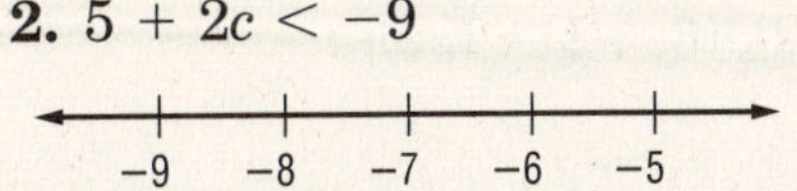

3. $4x - 3 < 2 - x$

4. $3(n + 2) < 24$

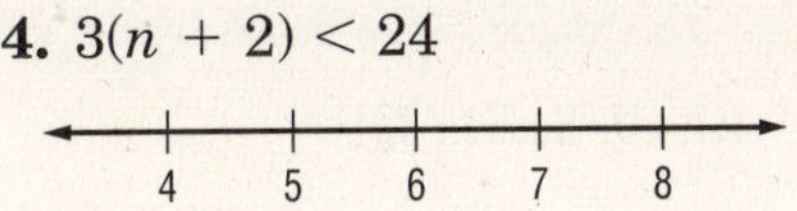

5. $11 + 2b \le 3(2 - b)$

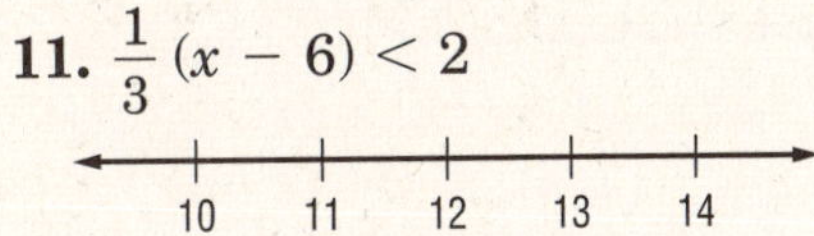

6. $\dfrac{m}{3} + 5 \ge 2$

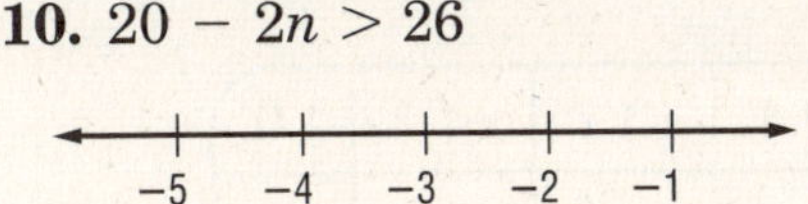

7. $\dfrac{1}{2}(8 - x) > 6$

8. $\dfrac{c}{4} + 7 \ge 5$

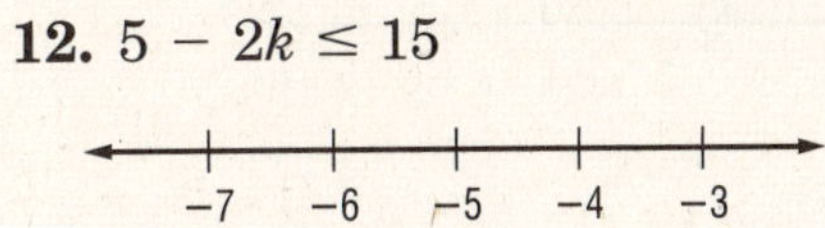

9. $y - 3 < 5y + 1$

10. $20 - 2n > 26$

11. $\dfrac{1}{3}(x - 6) < 2$

12. $5 - 2k \le 15$

13. $-2(3 + t) < -8$

14. $\dfrac{n}{4} - 9 > 5$

15. Two times a number less 4 is greater than the same number plus 6. For what number or numbers is this true?

16. One-half of the sum of a number and 4 is less than 14. What is the number?

17. **FISHING** Benjamin wants to go fishing on the lake. A boat rents for $12 per hour and a rod and reel rent for $20 per day. If he wants to spend no more than $80, how many hours can he spend fishing in the boat?

18. **ENTERTAINING** Deena is inviting 10 friends to a party. If she wants to spend no more than $120 on her guests, and dinner for each guest costs $8, what is the most can she spend on party favors for each person?

8-1 Skills Practice

Functions

Determine whether each relation is a function. Explain.

1. {(3, −8), (3, 2), (6, −1), (2, 2)}

2. {(0, 1), (−4, −3), (−3, 6), (3, 6)}

3. {(−6, 3), (2, −2), (0, 8), (1, 1)}

4. {(1, 8), (−6, 21), (−11, 21), (−3, 11), (0, 21)}

5.

x	1	−3	8	−8	20
y	2	6	6	5	11

6.

x	4	11	8	−13	−4
y	2	−4	1	2	20

7.

x	−1.2	1.1	1.7	−1.2	1.0
y	2.8	2.3	−2.4	2.3	2.6

8.

x	7	0	−6	1	−11
y	−1	4	8	8	14

9. 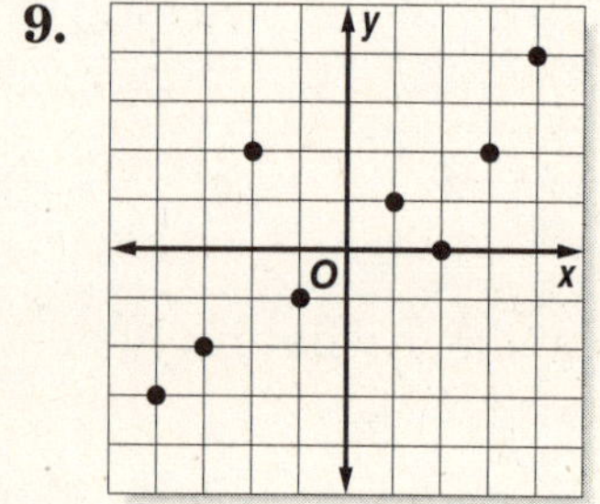

10. 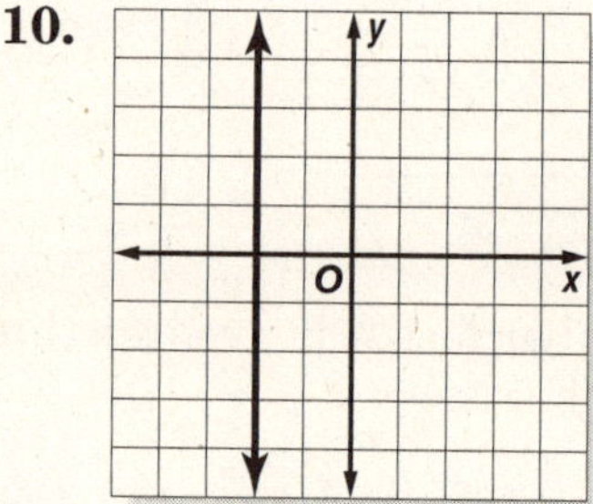

11. 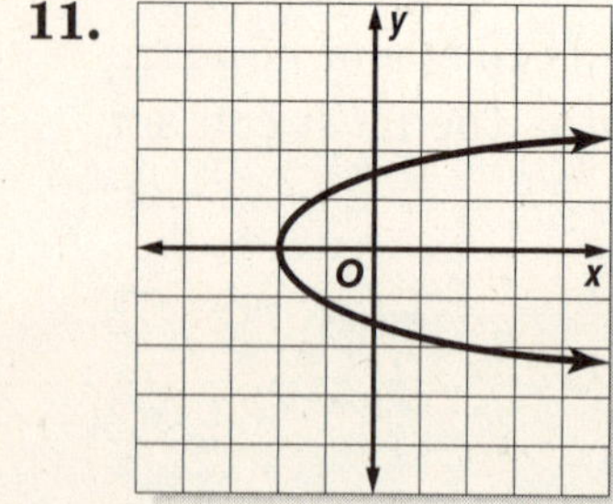

12. 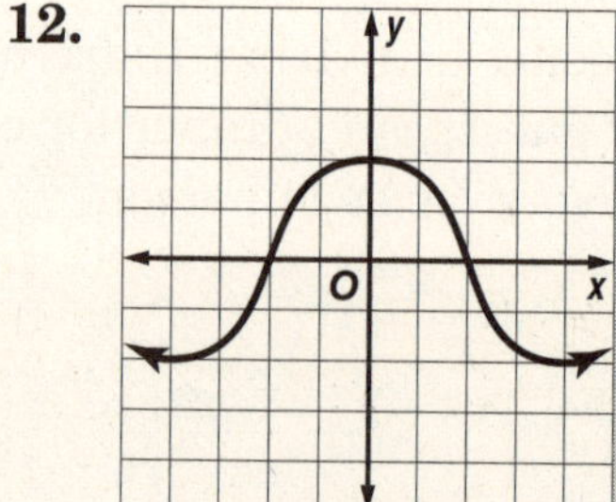

Glencoe Pre-Algebra

8-2 Skills Practice

Linear Equations in Two Variables

Find four solutions of each equation. Write the solutions as ordered pairs.

1. $y = 8x - 4$

2. $y = -x + 12$

3. $4x - 4y = 24$

4. $x - y = -15$

5. $y = 7x - 6$

6. $y = -3x + 8$

7. $y = 12$

8. $4x - 2y = 0$

9. $4x - y = 4$

Graph each equation by plotting ordered pairs.

10. $y = 3x - 2$

11. $y = -x + 3$

12. $y = -\dfrac{1}{2}x + \dfrac{3}{2}$

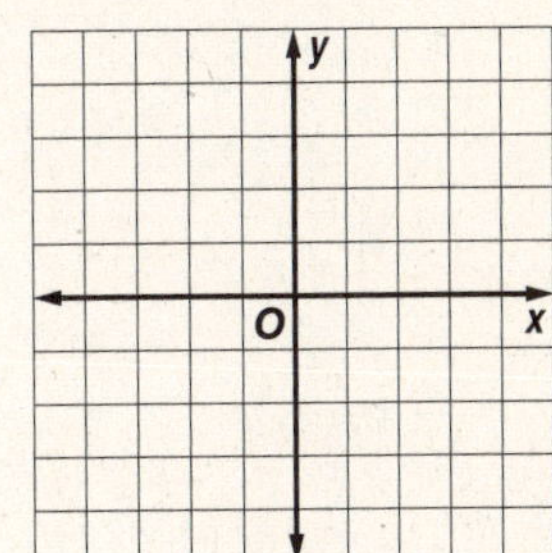
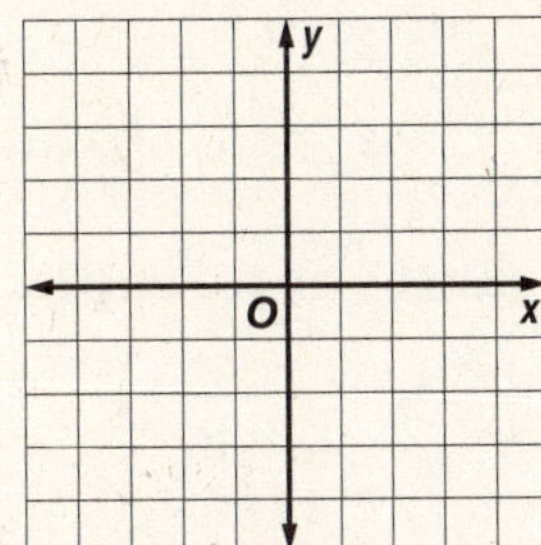
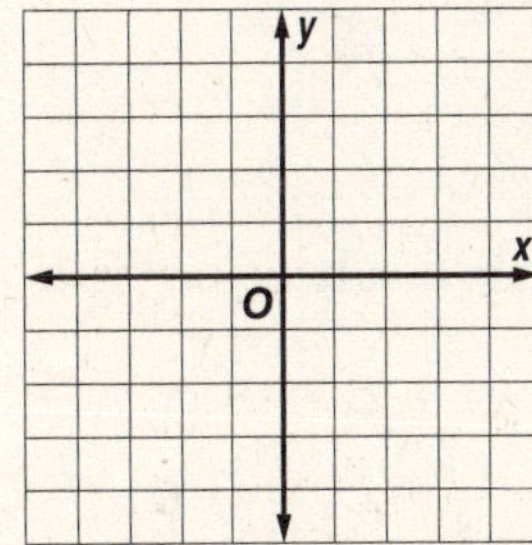

13. $y = -2x - 5$

14. $y = 4x - 8$

15. $y = \dfrac{2}{3}x - 2$

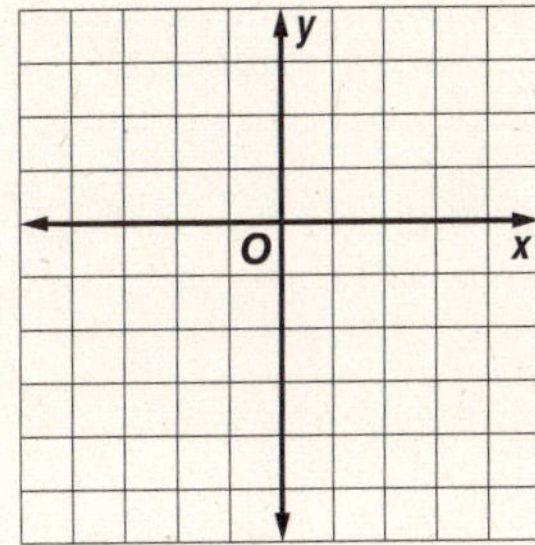
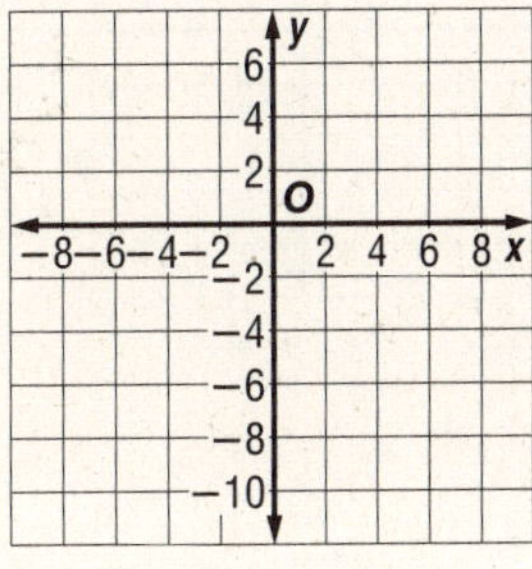
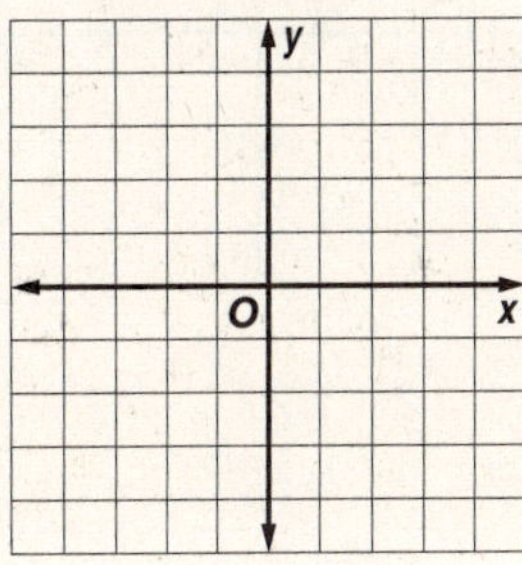

16. $y = -5x$

17. $y = -2x + 6$

18. $y = 5x + 1$

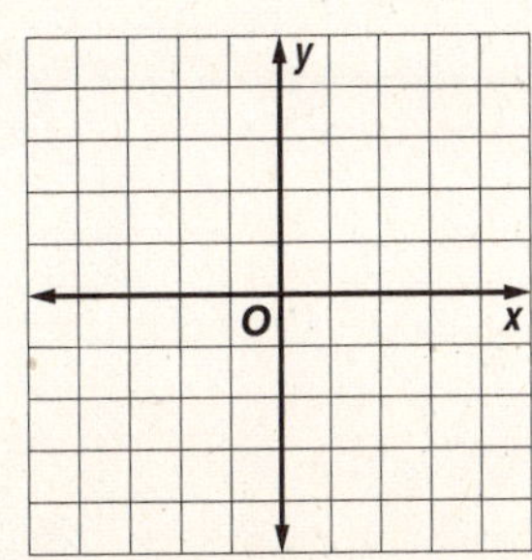
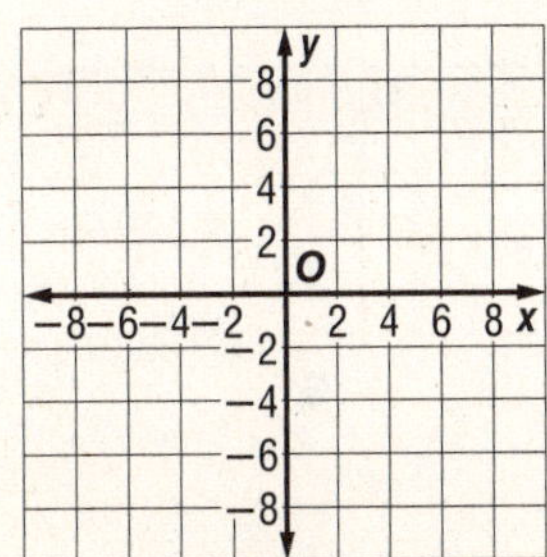
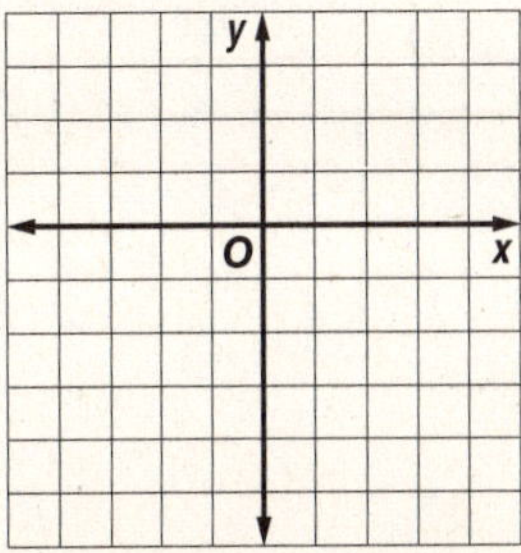

8-3 Skills Practice

Graphing Linear Equations Using Intercepts

State the *x*-intercept and the *y*-intercept of each line.

1. 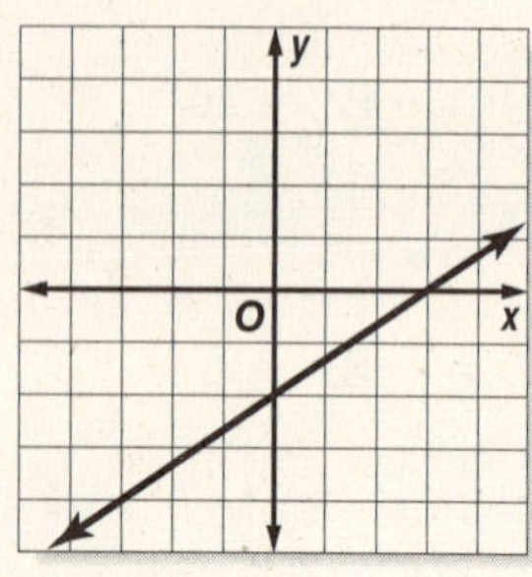

2. 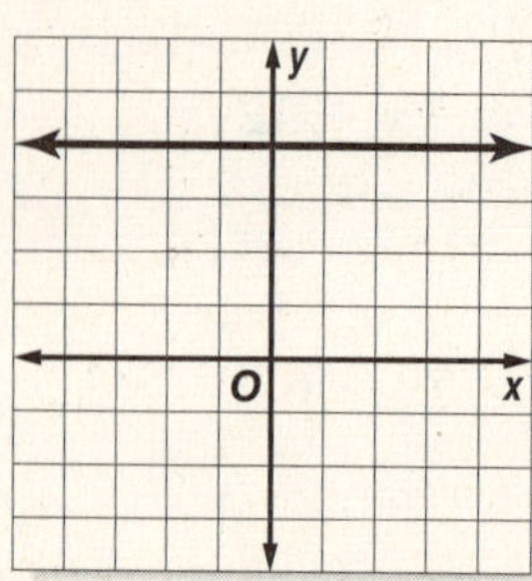

3. 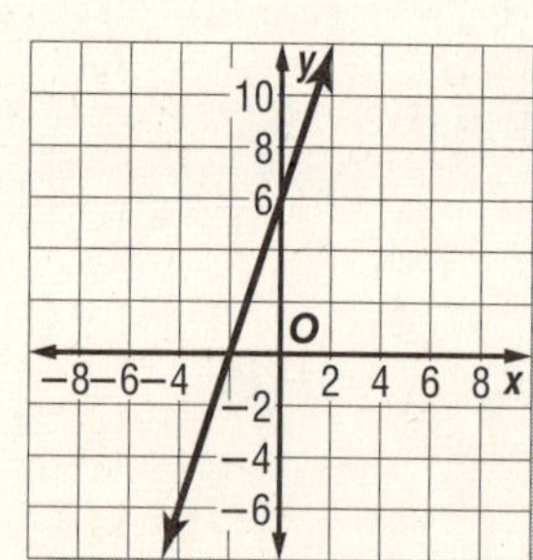

Find the *x*-intercept and the *y*-intercept for the graph of each equation.

4. $y = 2x + 6$

5. $3x - 5y = 30$

6. $y = -4x + 8$

7. $y = 7x - 14$

8. $y = 12x + 6$

9. $y = 7$

Graph each equation using the *x*- and *y*-intercepts.

10. $y = -2x + 6$

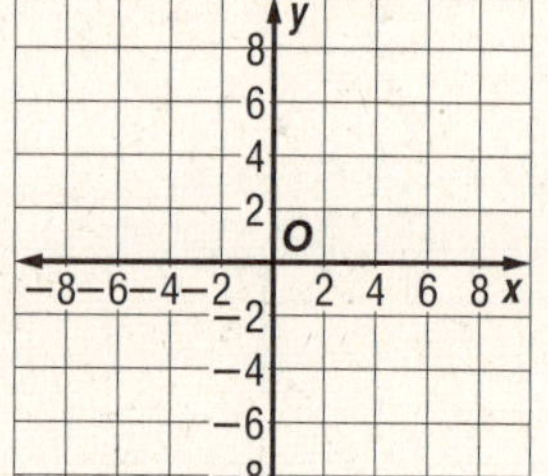

11. $y = -2$

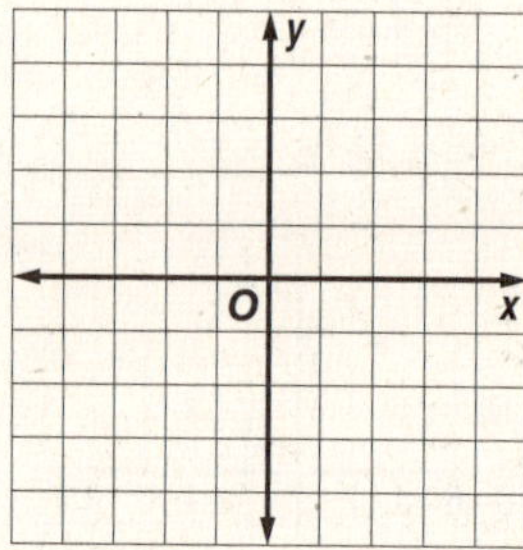

12. $y = -4x + 2$

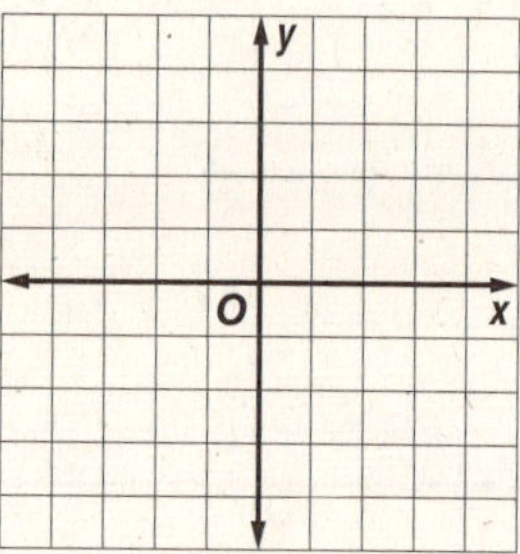

13. $y = \dfrac{2}{5}x - 2$

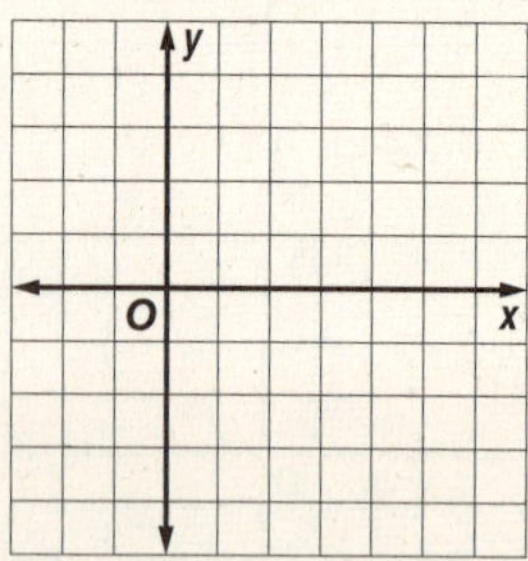

14. $x = 4$

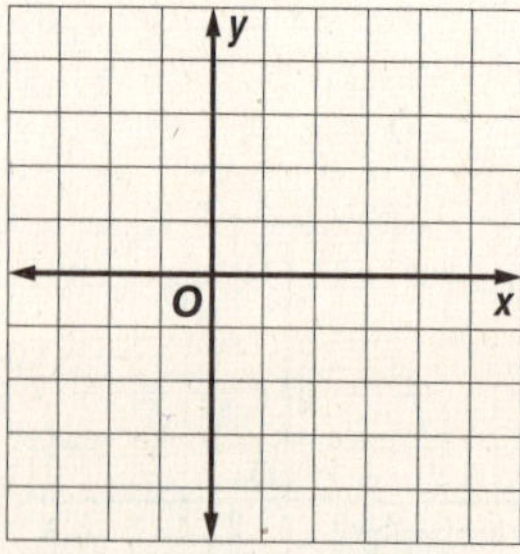

15. $y = -x + 3$

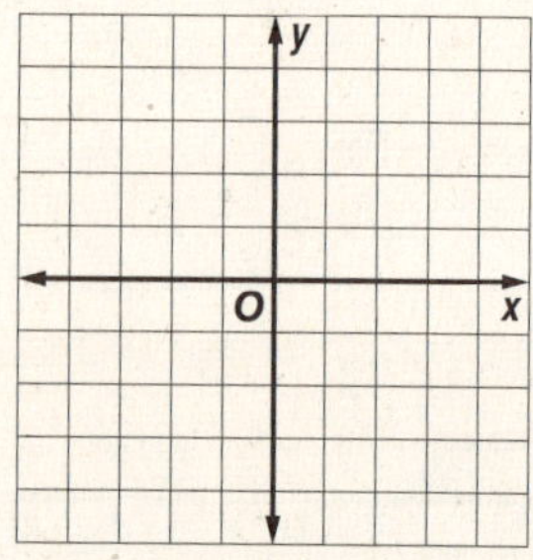

8-4 Skills Practice

Slope

Find the slope of each line.

1. 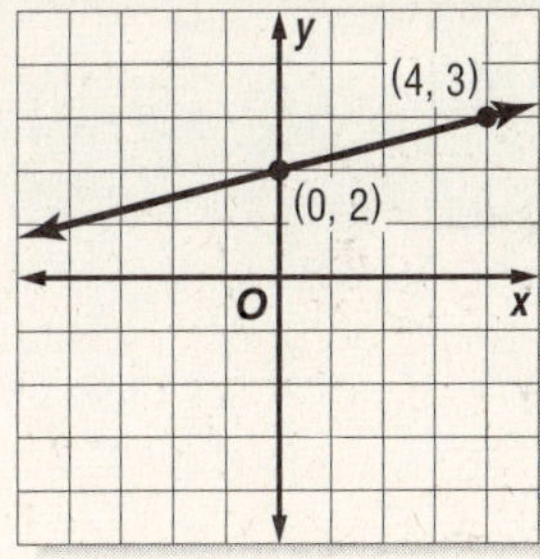

2. 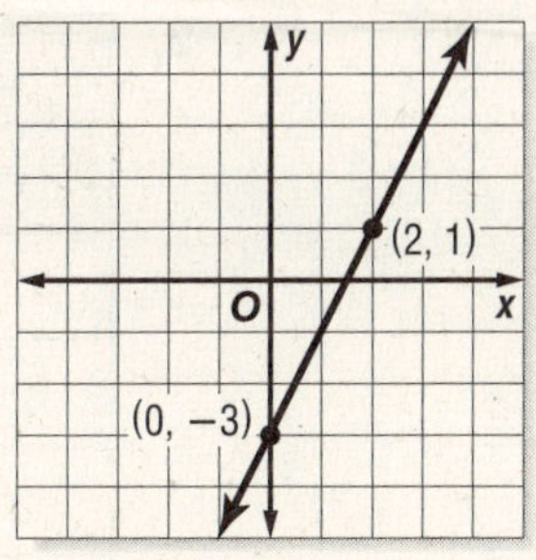

3. 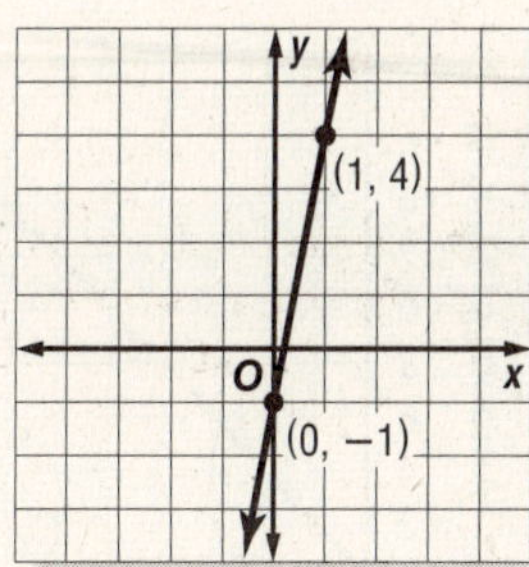

4. 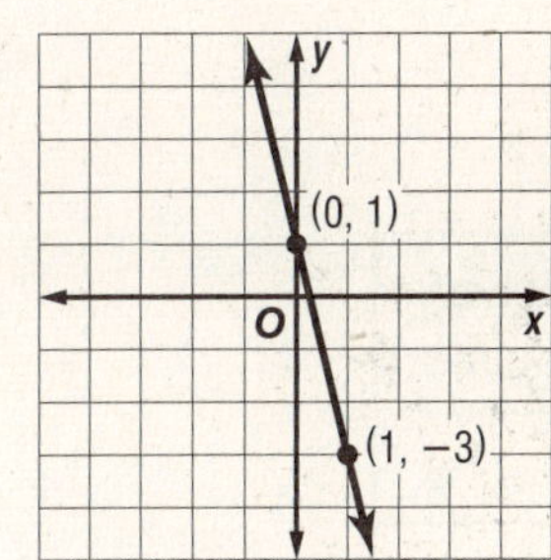

5. 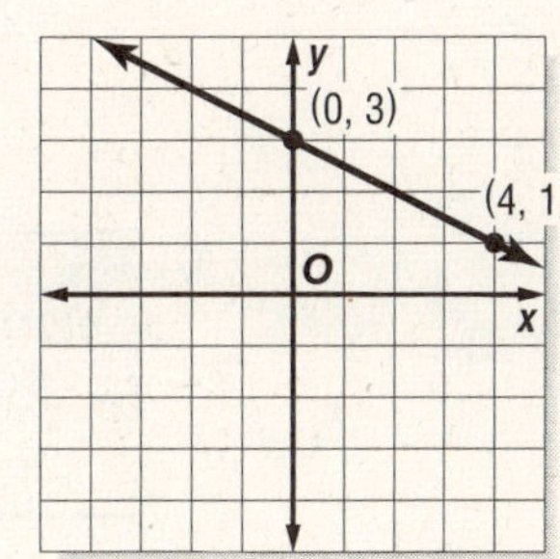

6. 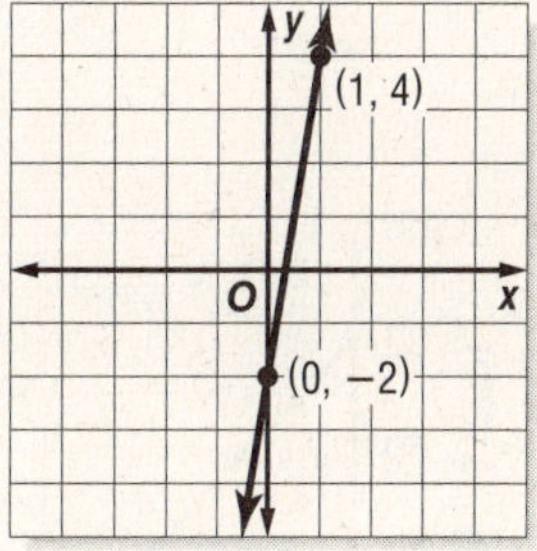

Find the slope of the line that passes through each pair of points.

7. $A(1, -5), B(6, -7)$

8. $C(7, -3), D(8, 1)$

9. $E(7, 2), F(12, 6)$

10. $G(8, -3), H(11, -2)$

11. $J(5, -9), K(0, -12)$

12. $L(-4, 6), M(5, 3)$

13. $P(2, -2), Q(7, -1)$

14. $R(-5, -2), S(-5, 3)$

15. $T(5, -6), U(8, -12)$

16. $P(10, -2), Q(3, -1)$

17. $R(6, -5), S(7, 3)$

18. $T(1, 8), U(7, 8)$

19. CAMPING   A family camping in a national forest builds a temporary shelter with a tarp and a 4-foot pole. The bottom of the pole is even with the ground, and one corner is staked 5 feet from the bottom of the pole. What is the slope of the tarp from that corner to the top of the pole?

20. ART   A rectangular painting on a gallery wall measures 7 meters high and 4 meters wide. What is the slope from the upper left corner to the lower right corner?

8-5 Skills Practice

Rate of Change

Find the rate of change for each linear function.

1.

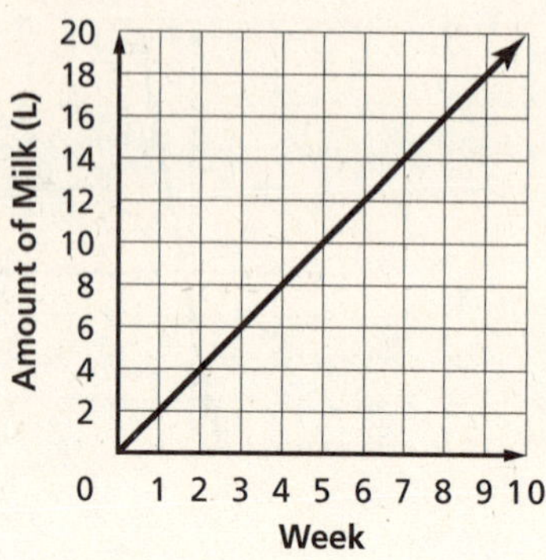

2.

Year	Salary ($)
x	y
1	21,000
2	23,500
3	26,000
4	28,500

3.

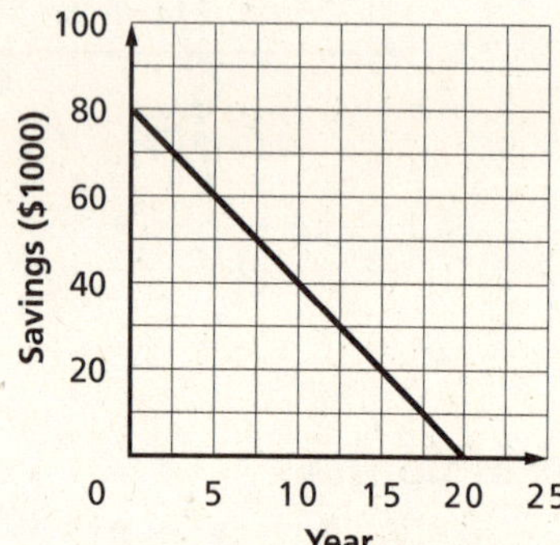

4.

Month	Number of Employees
x	y
0	0
2	22
4	44
6	66

Suppose y varies directly with x. Write an equation relating x and y.

5. $y = 2$ when $x = -7$　　　**6.** $y = 8$ when $x = -1$　　　**7.** $y = 1$ when $x = -2$

8. $y = 18$ when $x = 6$　　　**9.** $y = 5$ when $x = 25$　　　**10.** $y = 3$ when $x = 12$

11. $y = 1.5$ when $x = 5$　　　**12.** $y = 9$ when $x = -3$　　　**13.** $y = 1$ when $x = -5$

14. $y = 21$ when $x = 3$　　　**15.** $y = 54$ when $x = -6$　　　**16.** $y = 12$ when $x = -4$

8-6 Skills Practice

Slope-Intercept Form

State the slope and the *y*-intercept for the graph of each equation.

1. $y = 12x - 4$

2. $y = \frac{1}{4}x + 3$

3. $3x - y = 6$

Given the slope and *y*-intercept, graph each line.

4. slope $= -2$,
   y-intercept $= 2$

5. slope $= \frac{1}{2}$,
   y-intercept $= 4$

6. slope $= \frac{2}{3}$,
   y-intercept $= -3$

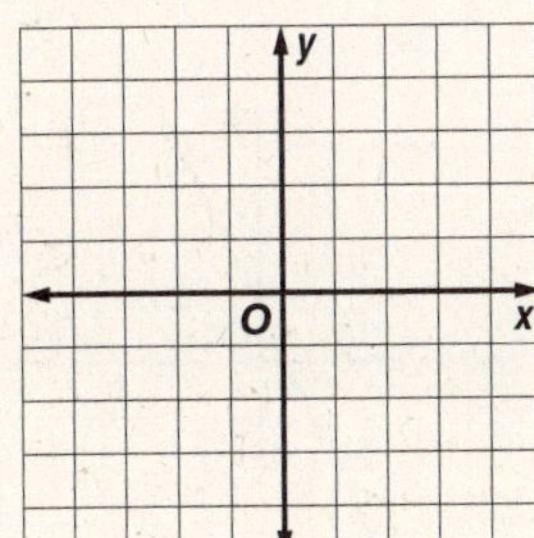 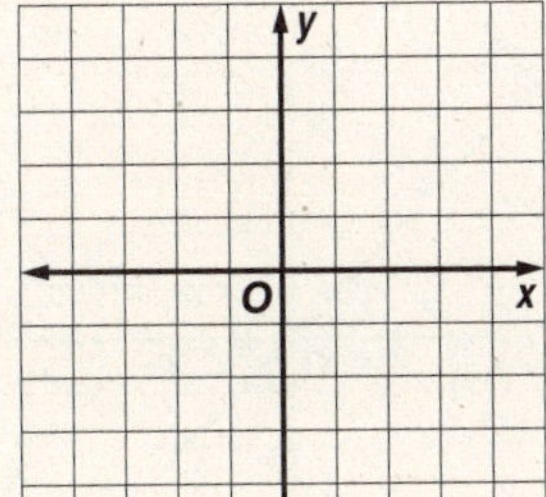 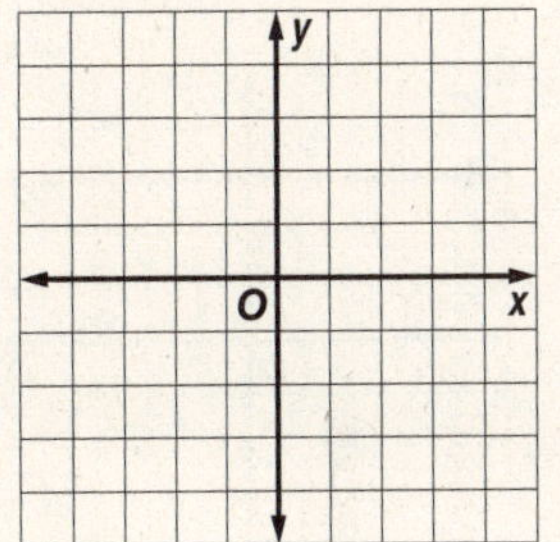

Graph each equation using the slope and *y*-intercept.

7. $y = 5x - 1$

8. $y = \frac{1}{2}x + 4$

9. $y = -x + 2$

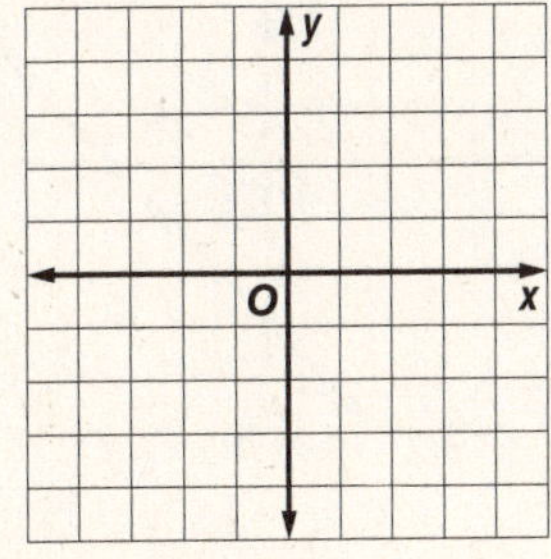 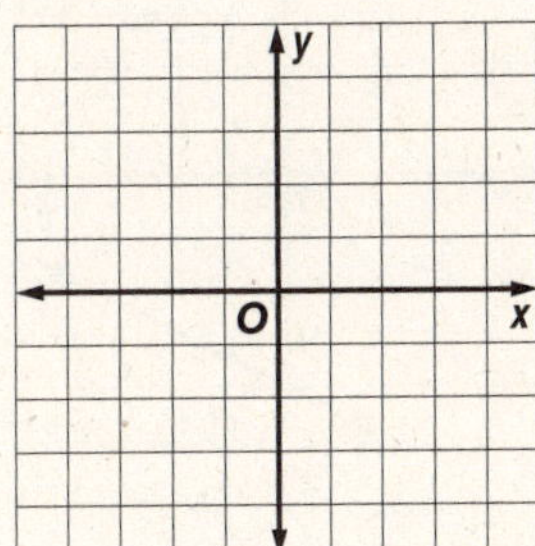 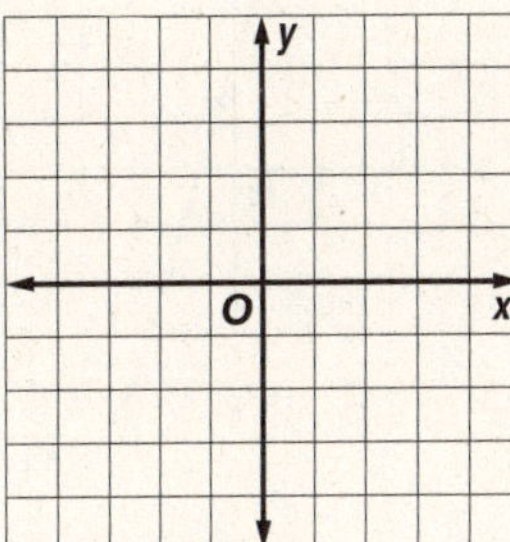

10. $y = 2x + 2$

11. $y = -4x + 2$

12. $y = x - 3$

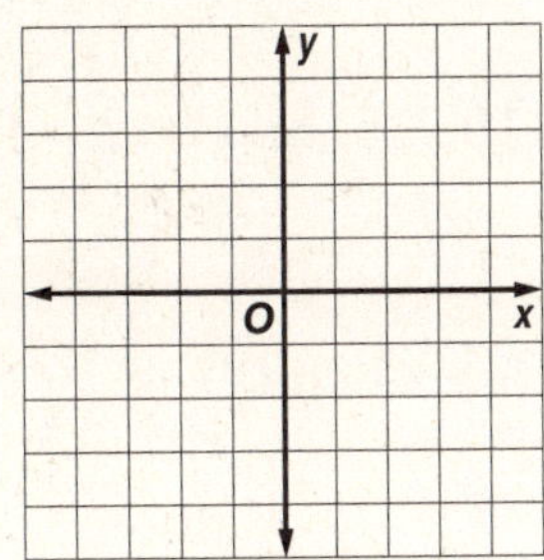 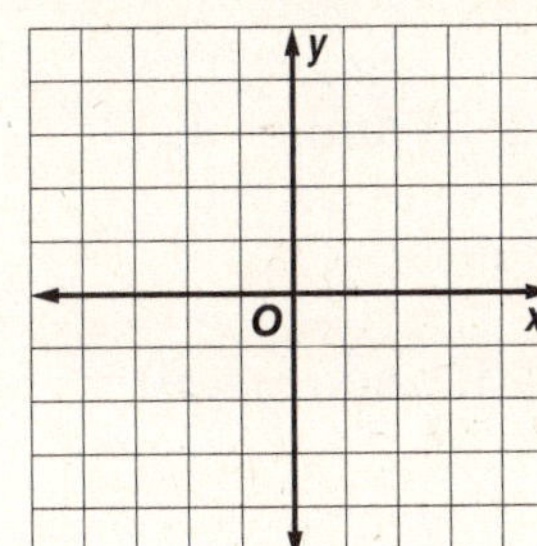 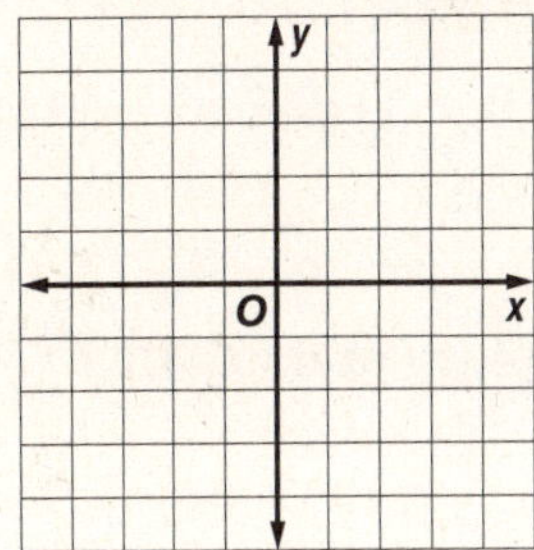

8-7 Skills Practice

Writing Linear Functions

Write an equation in slope-intercept form for each line.

1. slope = 7,
   y-intercept = 2

2. slope = −5,
   y-intercept = −3

3. slope = $\frac{3}{5}$,
   y-intercept = 6

4. slope = −6,
   y-intercept = 7

5. slope = $\frac{2}{7}$,
   y-intercept = 1

6. slope = $\frac{4}{3}$,
   y-intercept = −4

7. 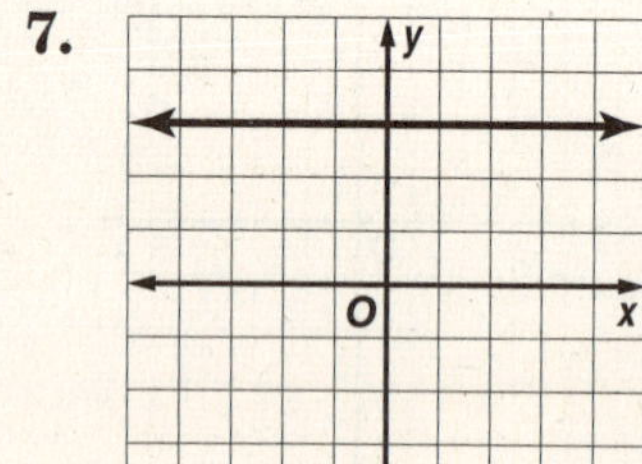

8. 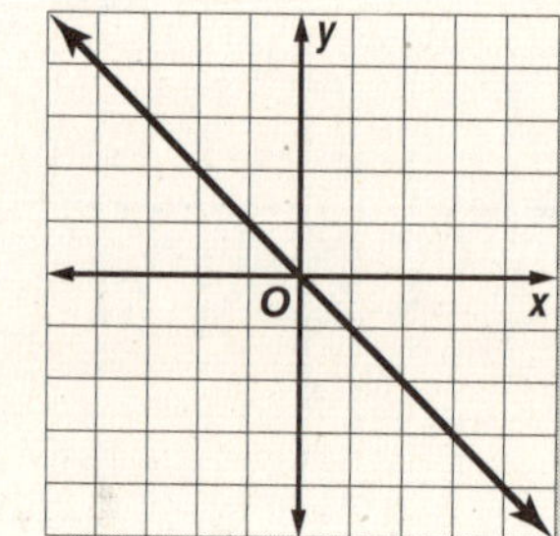

9. 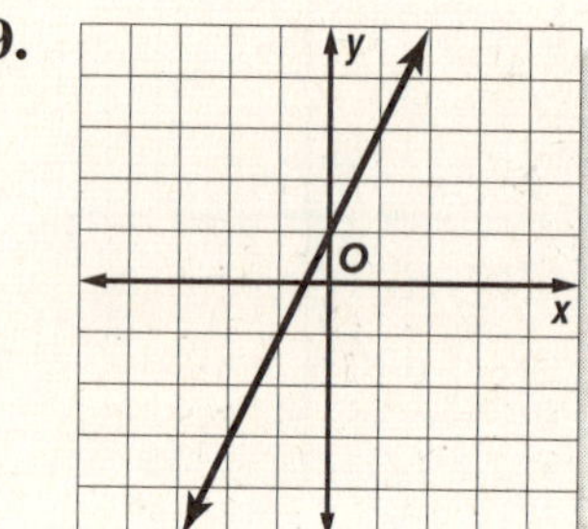

10. 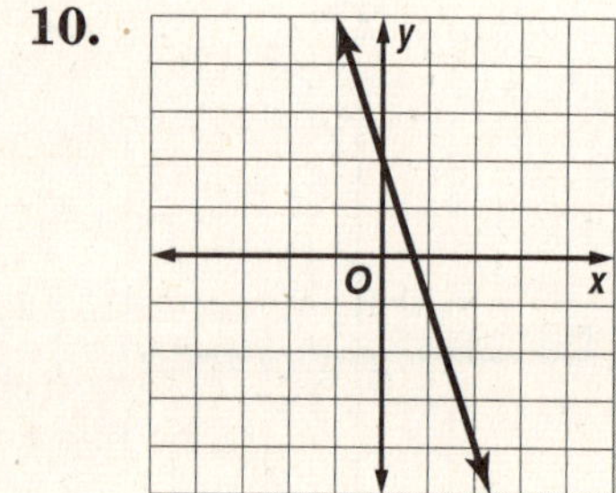

11. 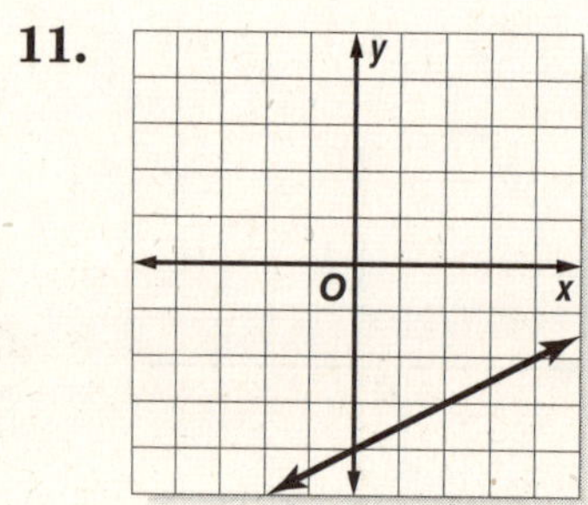

12. 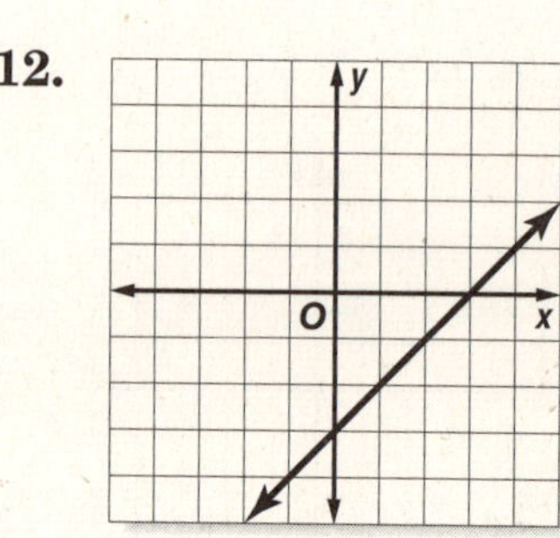

Write an equation in slope-intercept form for the line passing through each pair of points.

13. (9, −1) and (6, −2)

14. (12, 5) and (−4, 1)

15. (10, −6) and (−2, −6)

16. (4, 6) and (1, 3)

17. (6, 3) and (−6, 9)

18. (8, −4) and (−4, −1)

19. (5, 0) and (2, −3)

20. (12, −2) and (6, 2)

21. (−5, 10) and (3, −6)

Glencoe Pre-Algebra

8-8 Skills Practice

Best-Fit Lines

CONSTRUCTION For Exercises 1 and 2, use the table that shows the average hourly wage of U.S. construction workers from 1980 to 1999.

Year	Average Hourly Earnings ($)
1980	9.94
1985	12.32
1990	13.77
1995	15.09
1999	17.13

Source: U.S. Census Bureau

1. Make a scatter plot and draw a best-fit line.

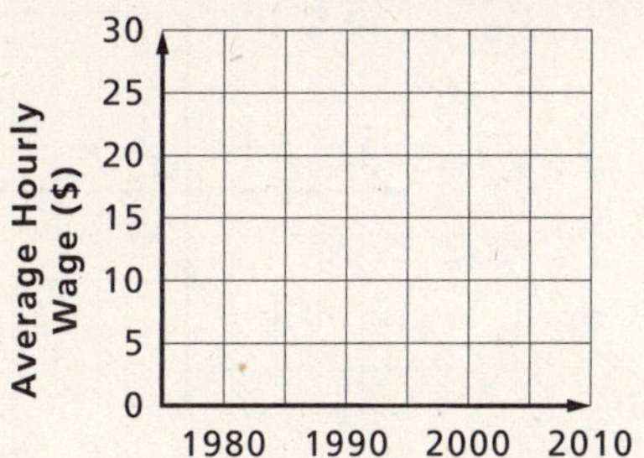

2. Use the best-fit line to predict the average hourly wage of construction workers in 2010.

MINING For Exercises 3 and 4, use the table that shows the number of persons employed in mining from 1980 to 1999.

Year	Employees (thousands)
1980	1027
1985	927
1990	709
1995	581
1999	535

Source: U.S. Census Bureau

3. Make a scatter plot and draw a best-fit line.

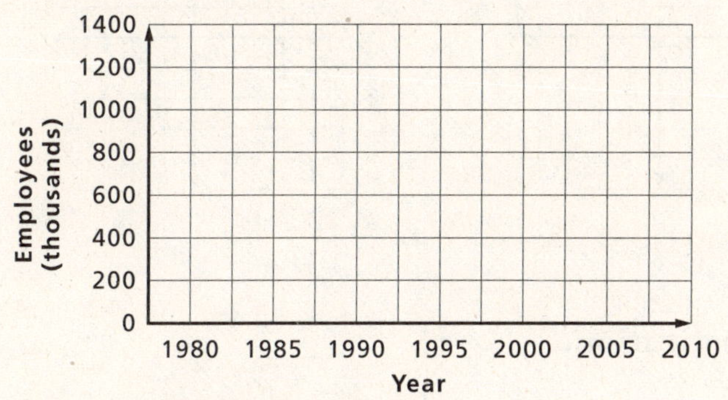

Source: U.S. Census Bureau

4. Write an equation for the best-fit line and use it to predict the number of persons employed in mining in 2010.

8-9 Skills Practice

Solving Systems of Equations

Solve each system of equations by graphing.

1. $y = 4x - 2$
$\quad y = -x + 3$

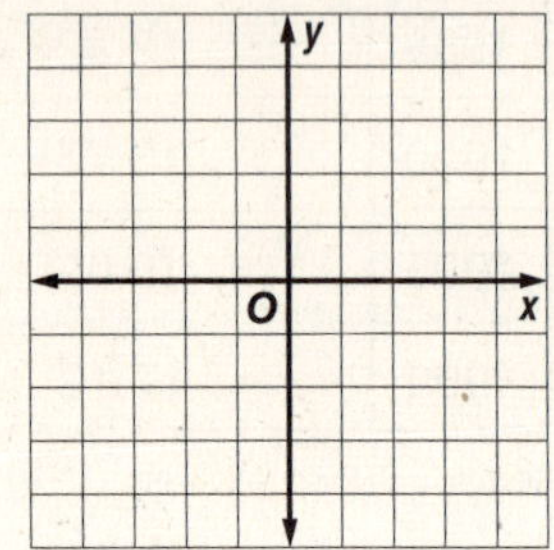

2. $y = -2x + 3$
$\quad 4x + 2y = -8$

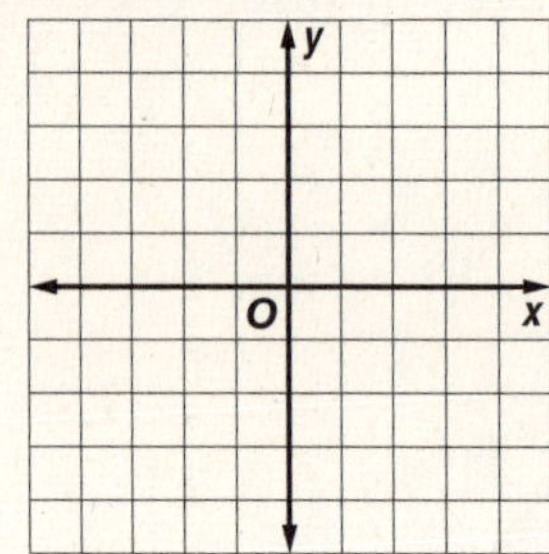

3. $y = -3x - 2$
$\quad y = x - 2$

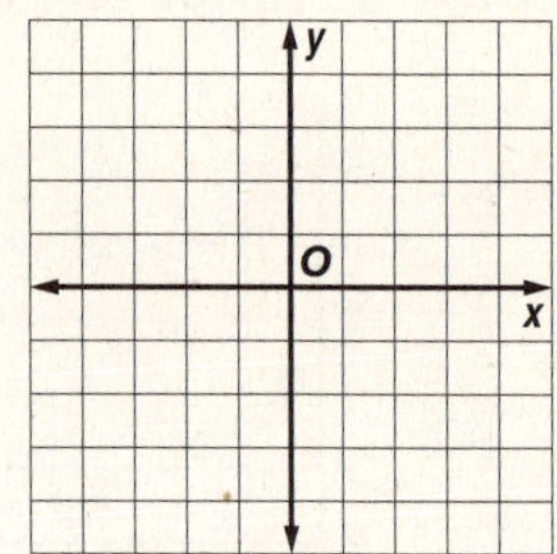

4. $x + y = 4$
$\quad y = 2x + 1$

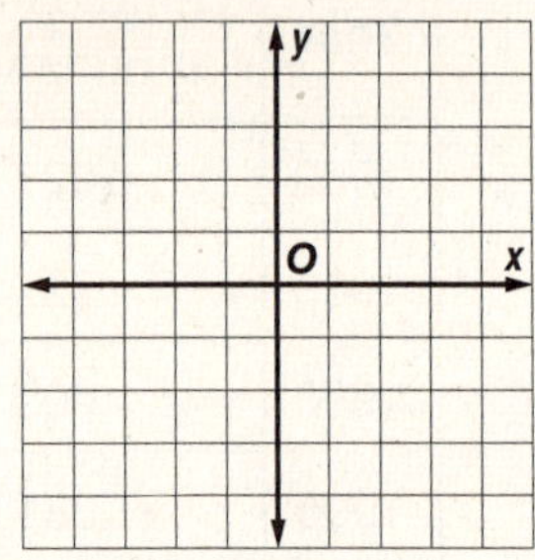

5. $3x + 4y = 12$
$\quad y = -\dfrac{3}{4}x + 3$

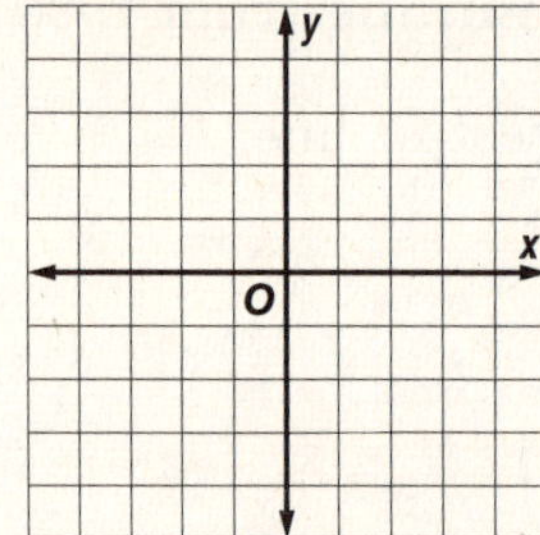

6. $y = -2x - 4$
$\quad y = x + 2$

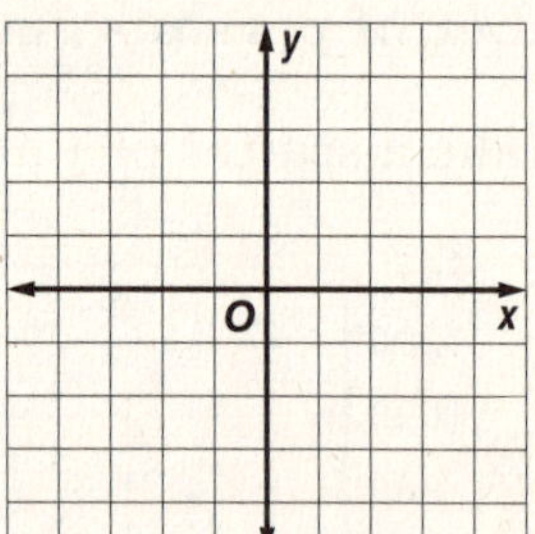

Solve each system of equations by substitution.

7. $y = 3x - 9$
$\quad y = 3$

8. $y = -x + 4$
$\quad x = 2$

9. $y = \dfrac{1}{2}x - 2$
$\quad y = 4$

10. $y = 6x$
$\quad x = \dfrac{1}{2}$

11. $2x + y = 8$
$\quad y = 2$

12. $x - y = 10$
$\quad y = 3$

8-10 Skills Practice

Graphing Inequalities

Graph each inequality.

1. $y > x + 2$

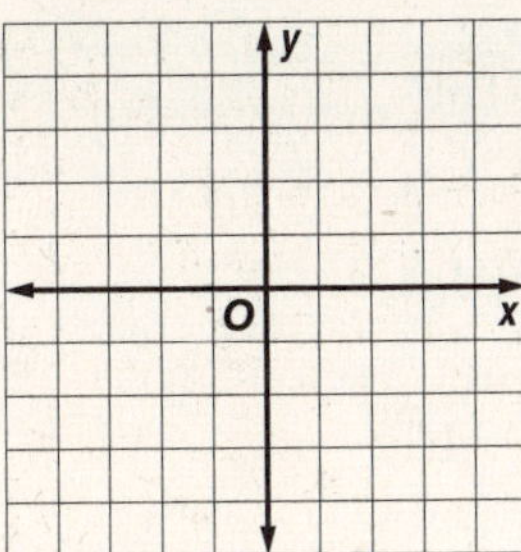

2. $y \leq x - 4$

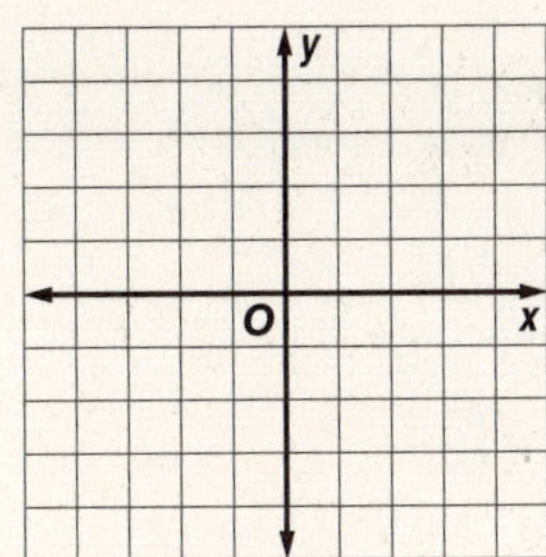

3. $y \geq \dfrac{1}{2}x$

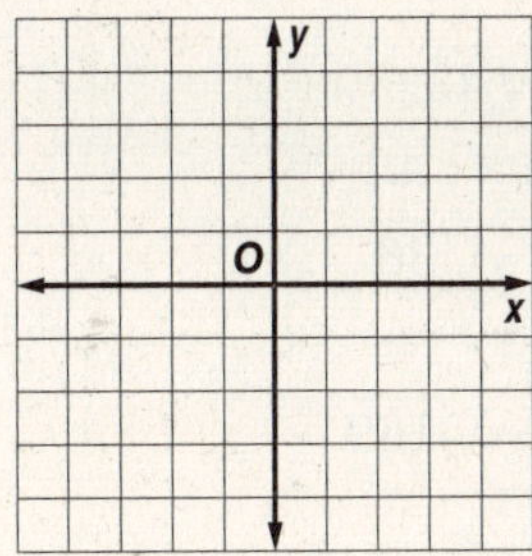

4. $y < 2x - 3$

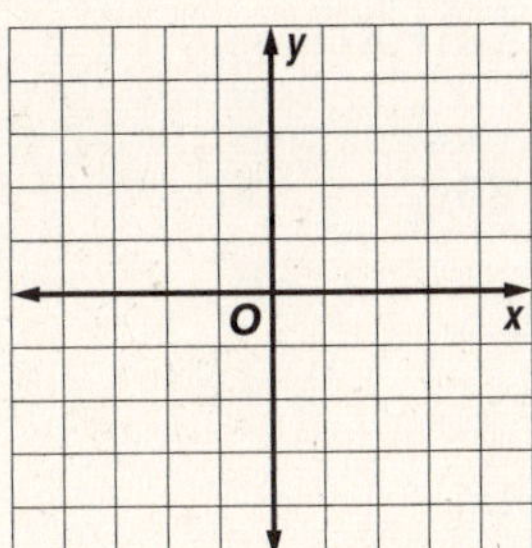

5. $y \geq -\dfrac{1}{3}x + 3$

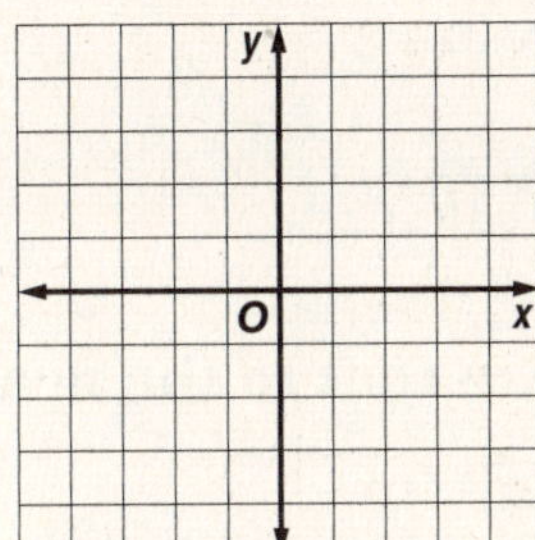

6. $y < 3x$

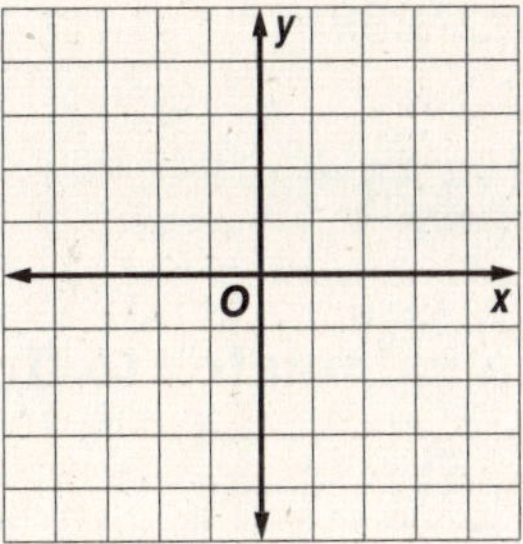

PACKING   For Exercises 7–9, use the following information.

The Gelbs are moving to a new home and want to pack efficiently. The boxes they are using can hold 10 pounds each. Suppose that each book weighs 1 pound and every sweater weighs 2 pounds. Assume that the boxes are large enough to hold any combination of books and sweaters weighing 10 pounds or less.

7. Write an inequality to represent this situation.

8. Graph the inequality.

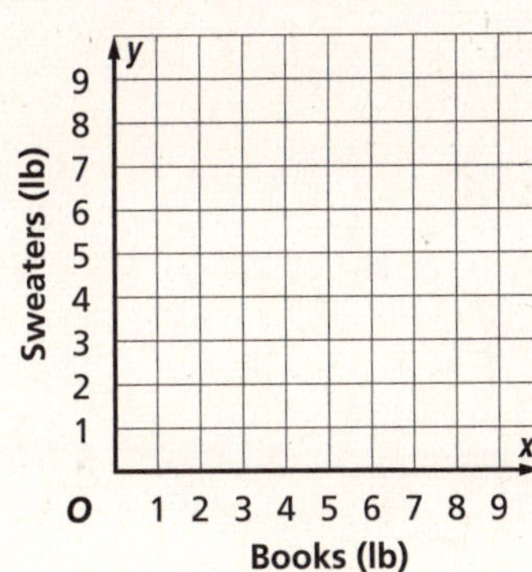

9. Use the graph to determine how many books and how many sweaters each box can hold.

9-1 Skills Practice

Squares and Square Roots

Find each square root, if possible.

1. $\sqrt{1}$ 2. $\sqrt{9}$ 3. $\sqrt{25}$

4. $\sqrt{49}$ 5. $\sqrt{64}$ 6. $\sqrt{169}$

7. $-\sqrt{36}$ 8. $\sqrt{-81}$ 9. $-\sqrt{64}$

10. $-\sqrt{169}$ 11. $\sqrt{-196}$ 12. $-\sqrt{121}$

13. $\sqrt{225}$ 14. $\sqrt{441}$ 15. $\sqrt{625}$

16. $\pm\sqrt{289}$ 17. $\pm\sqrt{324}$ 18. $\pm\sqrt{8100}$

19. $\sqrt{2.25}$ 20. $\sqrt{0.16}$ 21. $\sqrt{3.24}$

Use a calculator to find each square root to the nearest tenth.

22. $\sqrt{31}$ 23. $\sqrt{40}$ 24. $\sqrt{94}$

25. $\sqrt{132}$ 26. $-\sqrt{68}$ 27. $-\sqrt{247}$

28. $\sqrt{-521}$ 29. $-\sqrt{314}$ 30. $-\sqrt{902}$

31. $-\sqrt{0.85}$ 32. $-\sqrt{2.45}$ 33. $-\sqrt{4.05}$

Estimate each square root to the nearest whole number. Do not use a calculator.

34. $\sqrt{38}$ 35. $\sqrt{84}$ 36. $\sqrt{389}$

37. $\sqrt{5}$ 38. $\sqrt{118}$ 39. $\sqrt{230}$

40. $-\sqrt{83}$ 41. $-\sqrt{19}$ 42. $-\sqrt{119}$

43. $\sqrt{9.3}$ 44. $\sqrt{27.5}$ 45. $\sqrt{78.1}$

9-2 **Skills Practice**

The Real Number System

Name all of the sets of numbers to which each real number belongs. Let N = natural numbers, W = whole numbers, Z = integers, Q = rational numbers, and I = irrational numbers.

1. 12
2. 25
3. -5

4. $\dfrac{1}{8}$
5. $\dfrac{1}{9}$
6. $0.343434\ldots$

7. $\sqrt{31}$
8. $\sqrt{7}$
9. $\dfrac{25}{5}$

10. $-\dfrac{32}{4}$
11. 6.54
12. 24.6

13. 418
14. 0
15. $0.050050005\ldots$

Determine whether each statement is *sometimes*, *always*, or *never* true.

16. A whole number is a rational number.

17. A rational number is a natural number.

18. A negative number is an integer.

19. Zero is a natural number.

Replace each ● with <, >, or = to make a true statement.

20. $\sqrt{4}$ ● $2\dfrac{3}{7}$
21. $\sqrt{5}$ ● 2.1

22. $-\sqrt{12}$ ● -3.5
23. $\sqrt{104.04}$ ● 10.2

24. 7.8 ● $\sqrt{55}$
25. 15.1 ● $\sqrt{231}$

Order each set of numbers from least to greatest.

26. $5\dfrac{1}{3}$, 5.3, $\sqrt{28}$, $2\dfrac{1}{4}$

27. $\sqrt{53}$, $7\dfrac{1}{4}$, $\dfrac{36}{5}$, 7.27

28. -9.35, $-\sqrt{72.25}$, $-9\dfrac{2}{10}$, -9

ALGEBRA Solve each equation. Round to the nearest tenth, if necessary.

29. $a^2 = 64$
30. $d^2 = 169$
31. $f^2 = 441$

32. $76 = g^2$
33. $115 = h^2$
34. $k^2 = 450$

35. $b^2 = 4.41$
36. $y^2 = 0.36$
37. $m^2 = 0.0025$

9-3 **Skills Practice**

Angles

Use a protractor to find the measure of each angle. Then classify each angle as *acute, obtuse, right,* or *straight*.

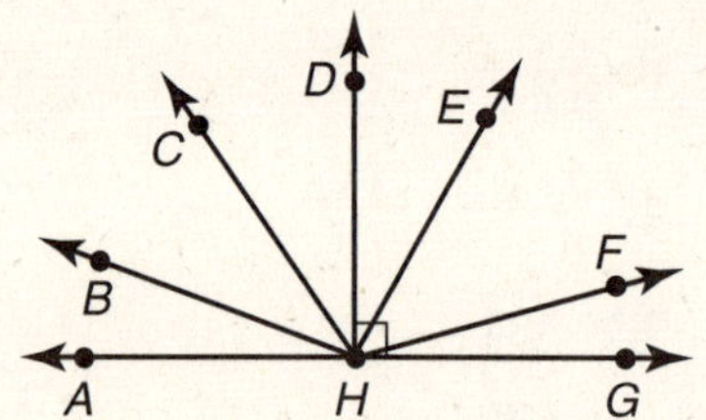

1. $\angle AHB$

2. $\angle AHC$

3. $\angle AHD$

4. $\angle AHE$

5. $\angle AHF$

6. $\angle AHG$

7. $\angle BHD$

8. $\angle DHG$

9. $\angle CHE$

10. $\angle CHF$

11. $\angle DHF$

12. $\angle BHF$

Use a protractor to draw an angle having each measurement. Then classify each angle as *acute, obtuse, right,* or *straight*.

13. $35°$

14. $45°$

15. $125°$

16. $170°$

17. $90°$

18. $180°$

19. $80°$

20. $135°$

21. $142°$

22. $87°$

9-4　**Skills Practice**

Triangles

Find the value of *x* in each triangle. Then classify each triangle as *acute*, *right*, or *obtuse*.

1. 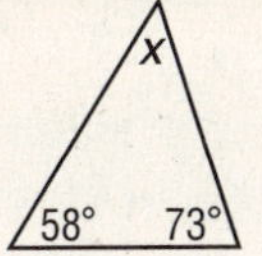

2. 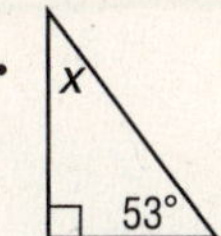

3. 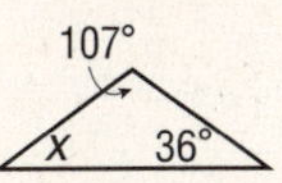

4. 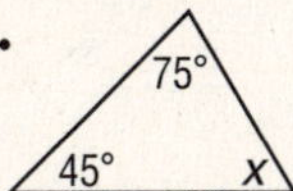

5. 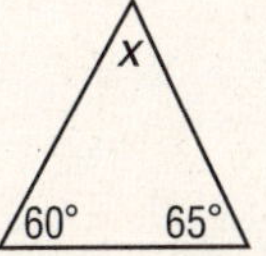

6. 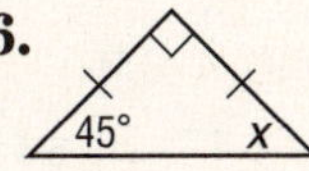

7.

8. 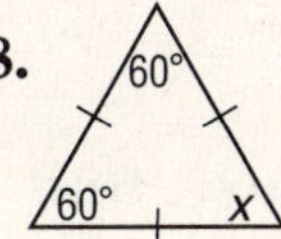

9. 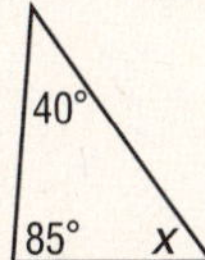

10. 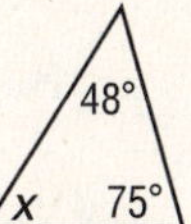

Classify each dashed triangle by its angles and by its sides.

11. 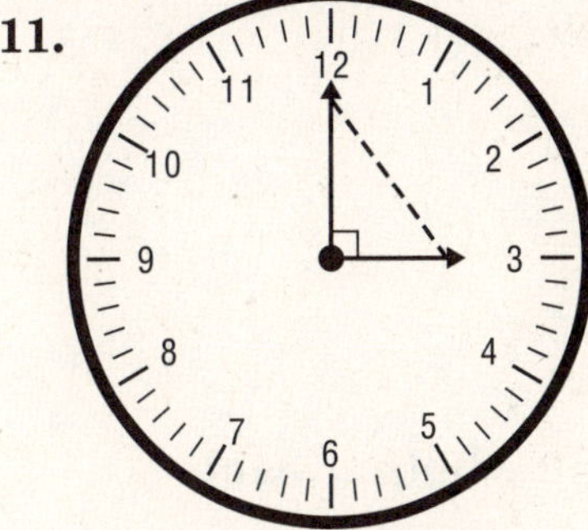

12. 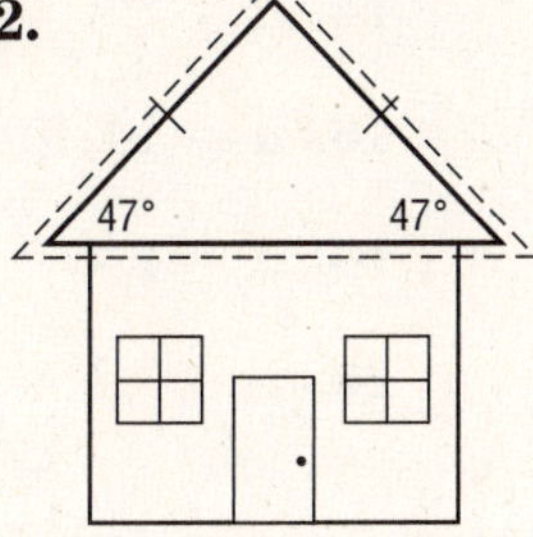

13. 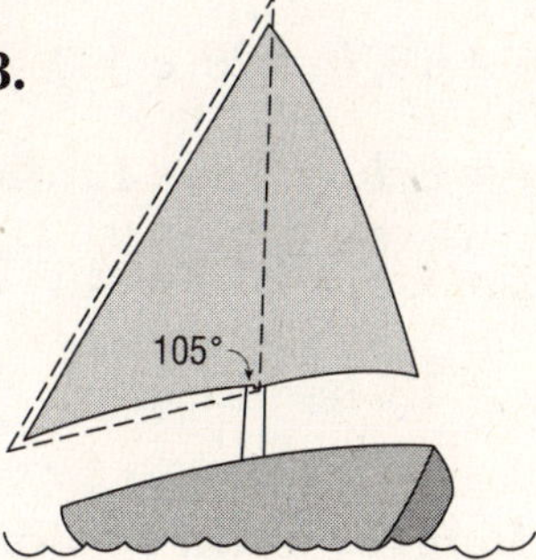

14. 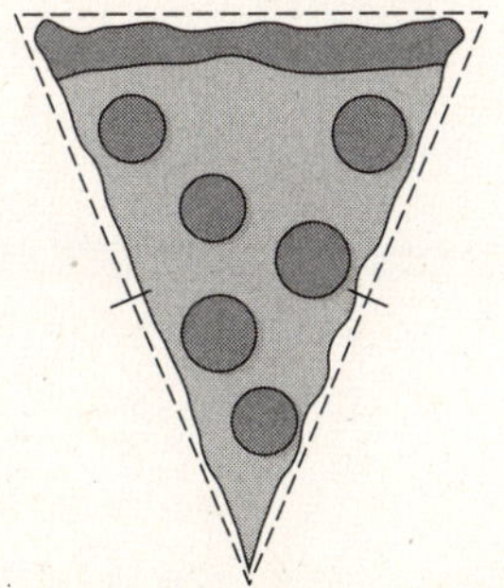

15. 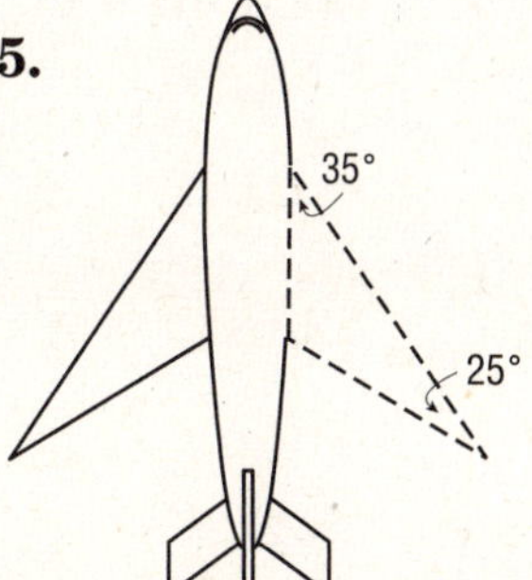

16. 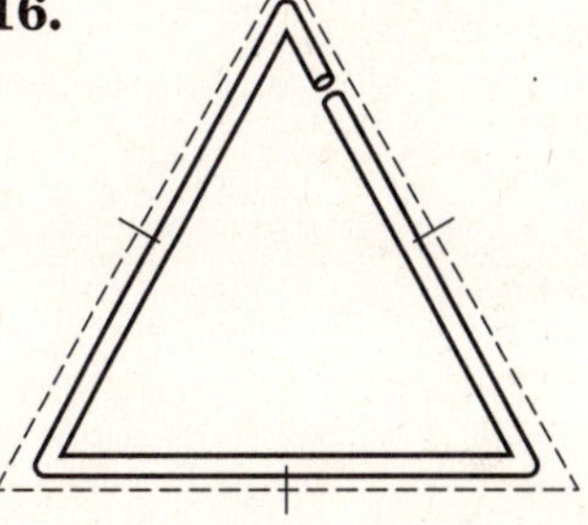

9-5 **Skills Practice**

The Pythagorean Theorem

Find the length of the hypotenuse in each right triangle. Round to the nearest tenth, if necessary.

1. 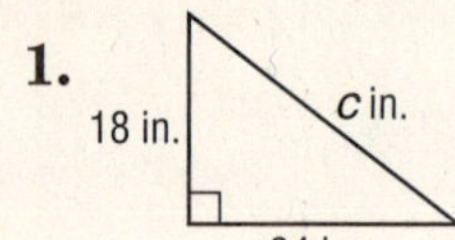

2. 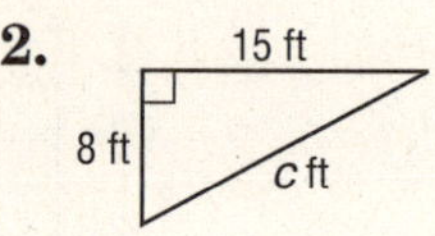

3. 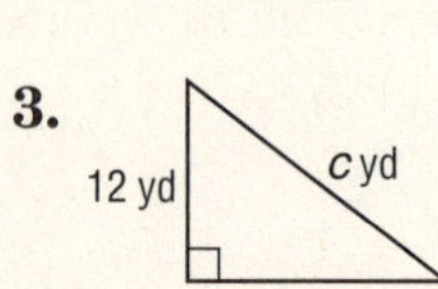

4. 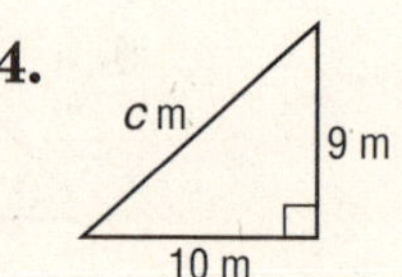

5. 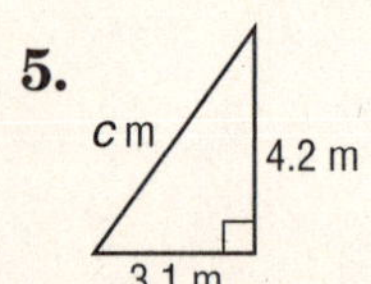

6. 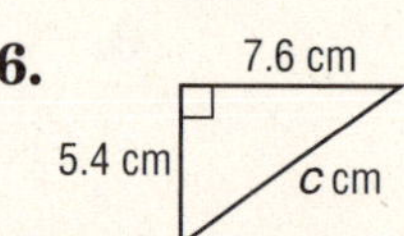

If c is the measure of the hypotenuse, find each missing measure. Round to the nearest tenth, if necessary.

7. $a = ?, b = 24, c = 26$

8. $a = 16, b = ?, c = 34$

9. $a = 24, b = ?, c = 40$

10. $a = 5, b = ?, c = 7$

11. $a = ?, b = 32, c = 39$

12. $a = 21, b = ?, c = 48$

13. $a = 18, b = 29, c = ?$

14. $a = ?, b = 36, c = 49$

15. $a = 8, b = ?, c = 12$

16. $a = 14, b = 21, c = ?$

17. $a = ?, b = 30, c = 40$

18. $a = 4, b = ?, c = 7$

19. $a = 13, b = 18, c = ?$

20. $a = ?, b = 55, c = 75$

The lengths of three sides of a triangle are given. Determine whether each triangle is a right triangle.

21. 14 m, 5 m, 4 m

22. 3 in., 4 in., 5 in.

9-6 Skills Practice

The Distance and Midpoint Formulas

Find the distance between each pair of points. Round to the nearest tenth, if necessary.

1. $A(2, 4), B(1, 3)$

2. $P(5, 10), Q(-1, 1)$

3. $G(3, -1), H(5, 6)$

4. $C(-2, -6), D(-7, 1)$

5. $E(-6, 2), F(4, 1)$

6. $J(-5, -3), K(4, -2)$

7. $M(-5, -5), N(3, -4)$

8. $V(4, 7), W(1, 6)$

9. $X(4, 6), Y(-3, -7)$

10. $R(0, 0), S(-1, -1)$

11. $T(7, 3), U(-2, -2)$

12. $A(6, 2), B(1, 3)$

GEOMETRY Find the perimeter of each figure.

13. 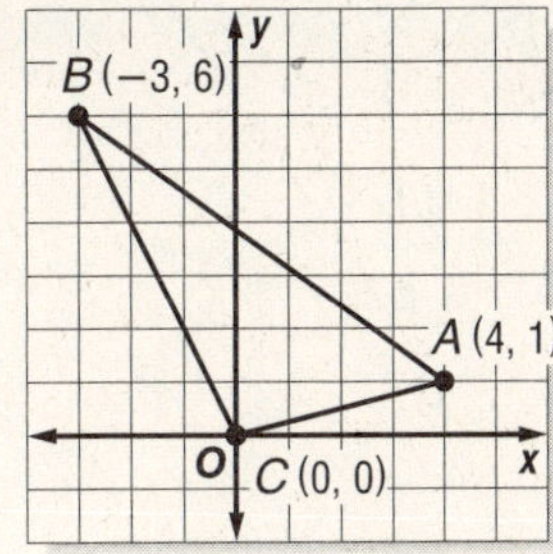

14. 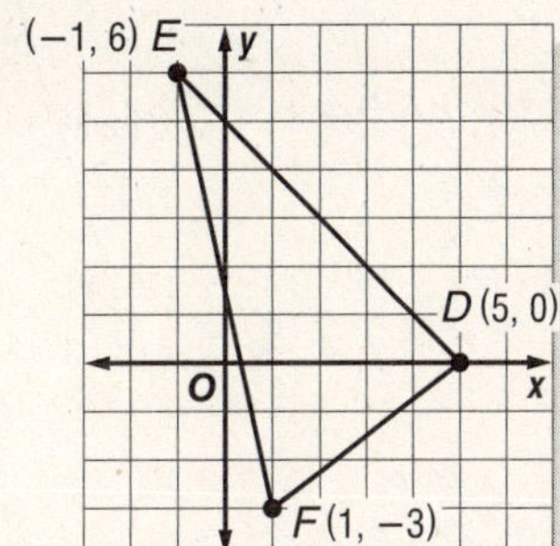

The coordinates of the endpoints of a segment are given. Find the coordinates of the midpoint of each segment.

15. $A(-5, -5), B(3, 5)$

16. $V(2, -6), W(4, -7)$

17. $C(6, 2), D(4, 7)$

18. $X(7, 8), Y(-7, 1)$

19. $E(7, 3), F(-1, 4)$

20. $A(5, 10), B(-4, -3)$

21. $G(-6, 2), H(2, 4)$

22. $C(-6, 7), D(-1, -1)$

23. $J(4, 1), K(-2, -2)$

24. $E(-4, 4), F(3, 5)$

25. $M(4, 8), N(-3, 4)$

26. $G(3, -1), H(5, 6)$

9-7 **Skills Practice**

Similar Triangles and Indirect Measurement

In Exercises 1–10, the triangles are similar. Write a proportion to find each missing measure. Then find the value of x.

1. 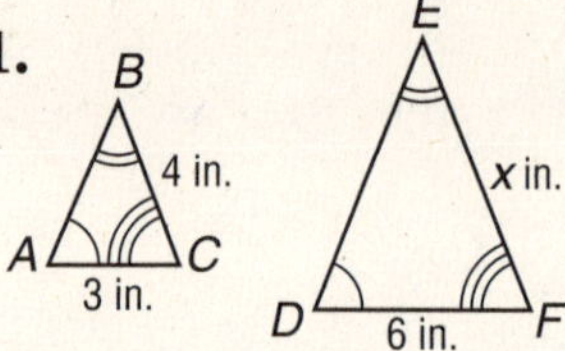

2. 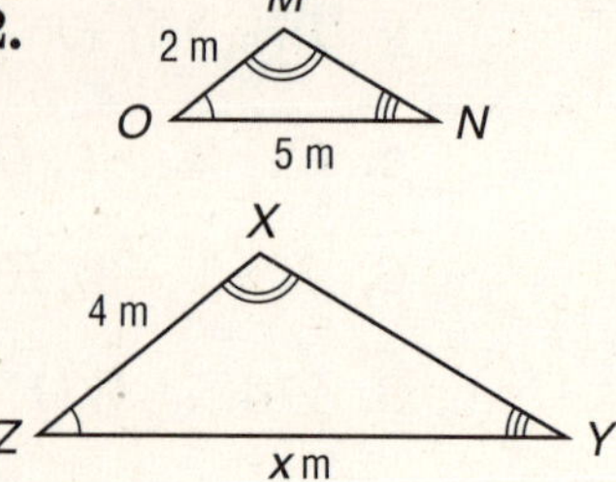

3. 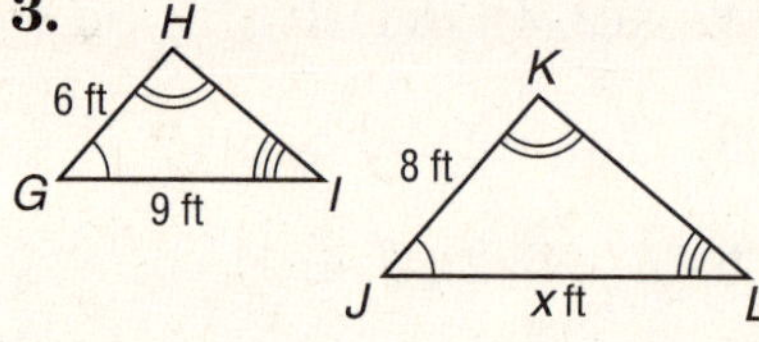

4. 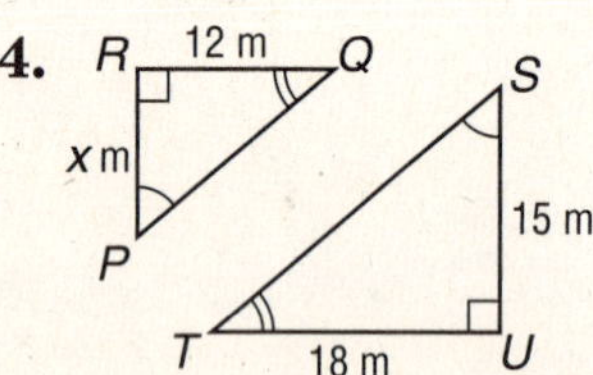

5. 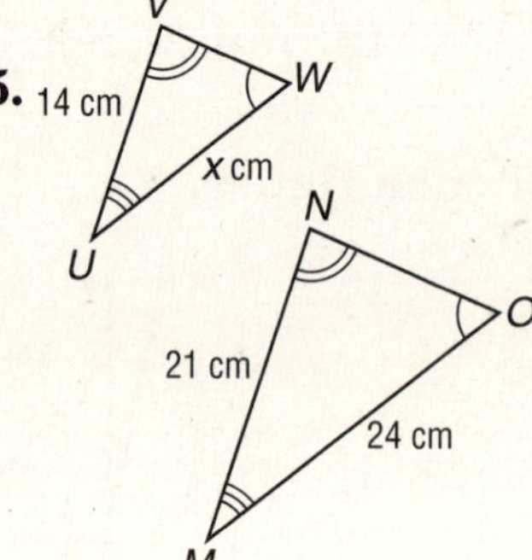

6. 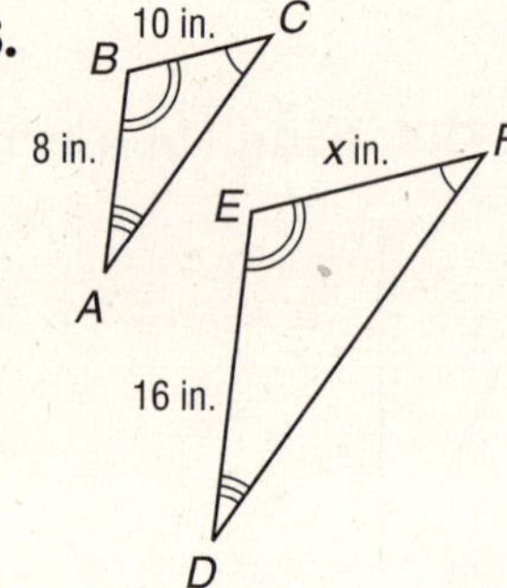

7. 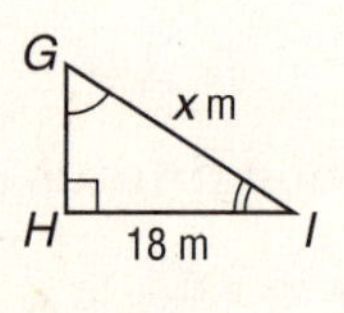

8. 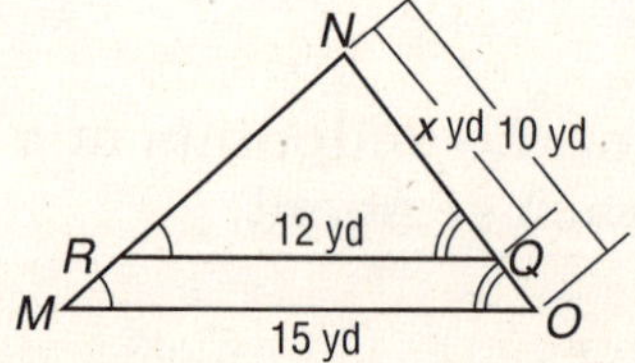

9. How far is the store from the bank?

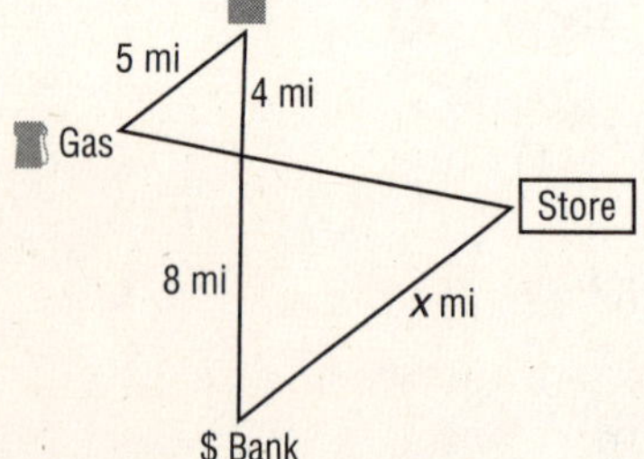

10. How far is the tree from the flagpole?

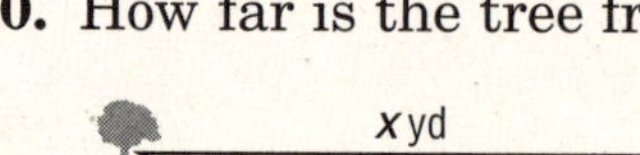
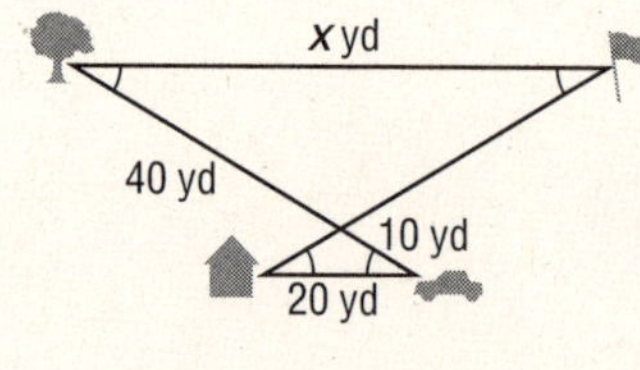

For Exercises 11 and 12, write a proportion. Then determine the missing measure.

11. ANIMALS At the same time a 12-foot adult elephant casts a 4.8-foot shadow, a baby elephant casts a 2-foot shadow. How tall is the baby elephant?

12. AIRPORTS If a 12-meter-tall airplane hangar casts a 18-meter shadow at the same time a parked jet casts a 6-meter shadow, how tall is the jet?

9-8 Skills Practice

Sine, Cosine, and Tangent Ratios

Find each sine, cosine, or tangent. Round to four decimal places, if necessary.

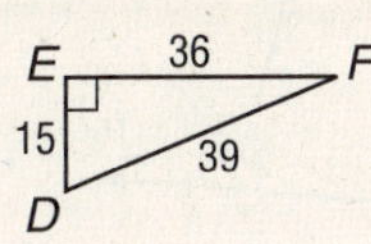

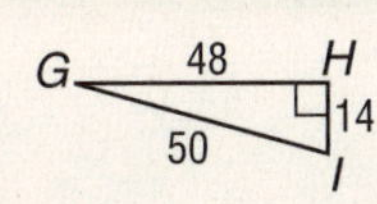

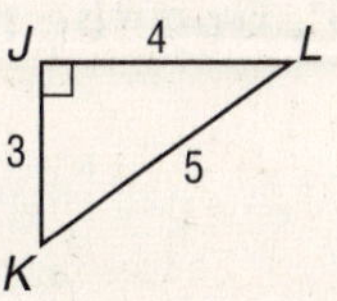

1. $\sin D$	**2.** $\sin G$	**3.** $\sin K$
4. $\cos D$	**5.** $\cos G$	**6.** $\cos K$
7. $\tan D$	**8.** $\tan G$	**9.** $\tan K$
10. $\sin F$	**11.** $\sin I$	**12.** $\sin L$
13. $\cos F$	**14.** $\cos I$	**15.** $\cos L$
16. $\tan F$	**17.** $\tan I$	**18.** $\tan L$

Use a calculator to find each value to the nearest ten thousandth.

19. $\sin 10°$	**20.** $\tan 25°$	**21.** $\cos 15°$
22. $\cos 52°$	**23.** $\sin 90°$	**24.** $\tan 53°$
25. $\tan 48°$	**26.** $\cos 0°$	**27.** $\sin 58°$
28. $\sin 89°$	**29.** $\tan 0°$	**30.** $\cos 16°$
31. $\cos 90°$	**32.** $\sin 0°$	**33.** $\tan 5°$

For each triangle, find each missing measure to the nearest tenth.

34. 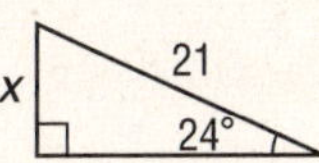

35. 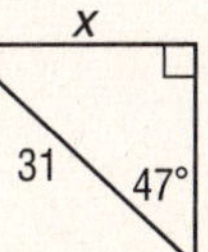

36. 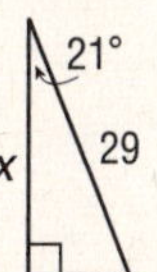

37. 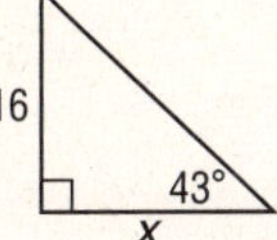

38. 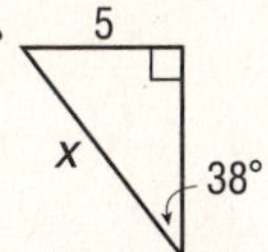

39. 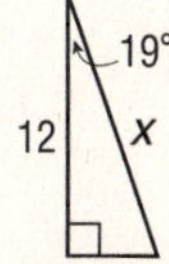

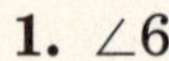

10-1 Skills Practice

Line and Angle Relationships

In the figure at the right, $c \parallel d$ and p is a transversal.
If $m\angle 5 = 110°$, find the measure of each angle.

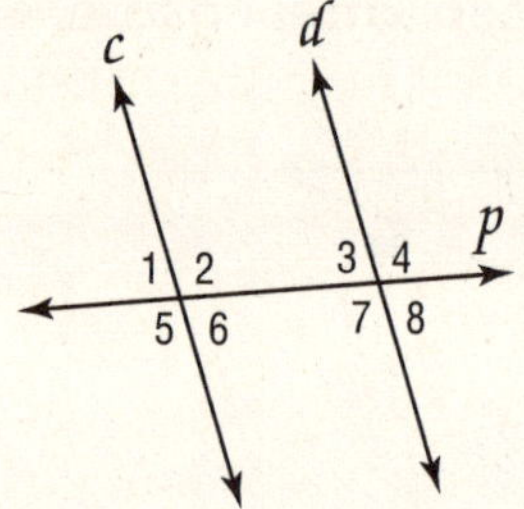

1. $\angle 6$

2. $\angle 8$

3. $\angle 2$

4. $\angle 4$

In the figure at the right, $g \parallel k$ and r is a transversal.
If $m\angle 7 = 60°$, find the measure of each angle.

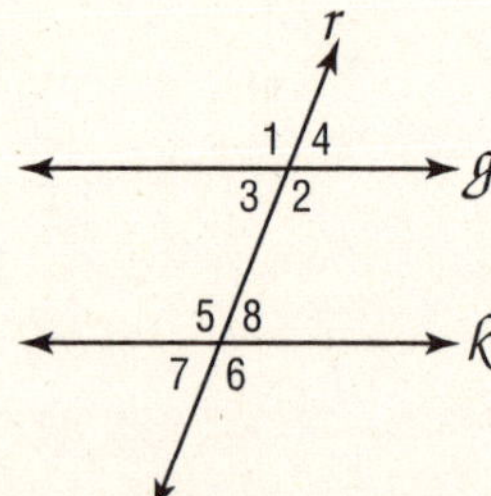

5. $\angle 4$

6. $\angle 6$

7. $\angle 5$

8. $\angle 3$

Find the value of x in each figure.

9.
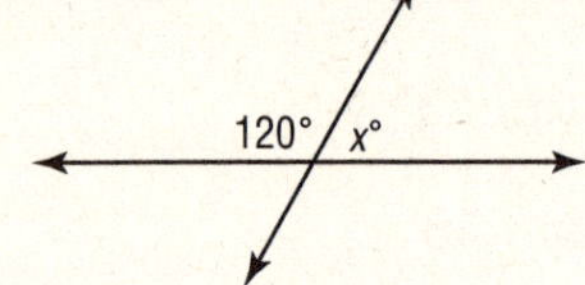

10.
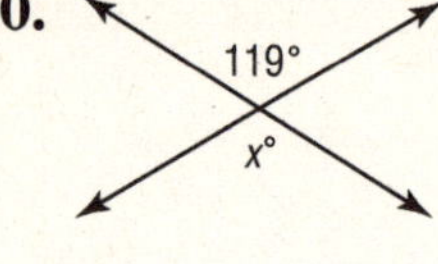

11.
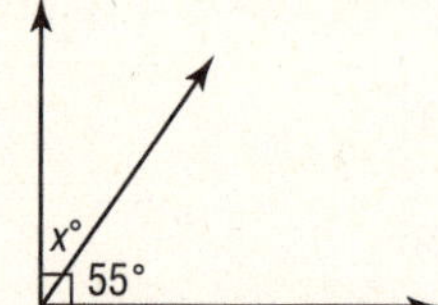

12.
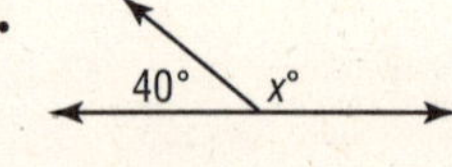

13.
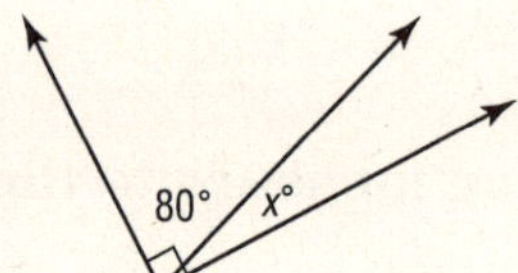

14.
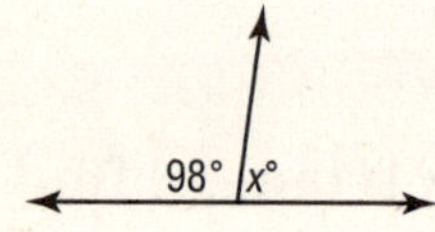

15.
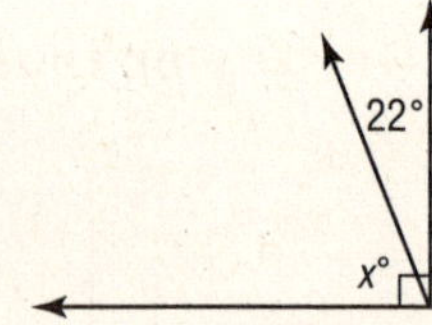

16.
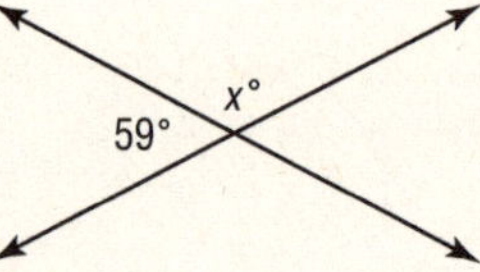

17.
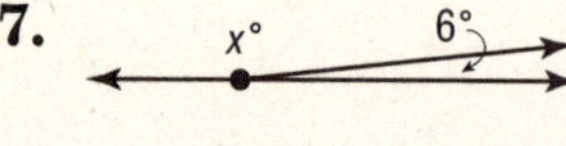

18.
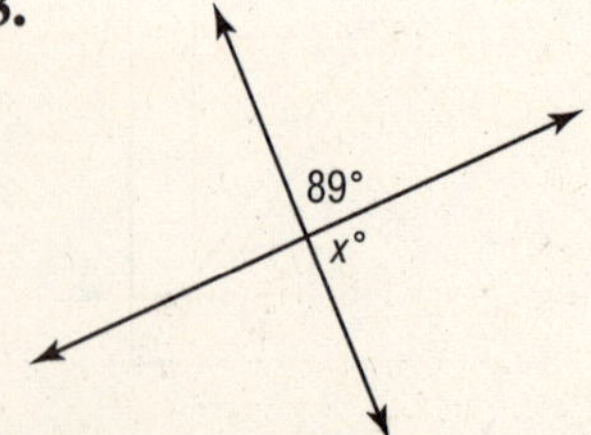

19.
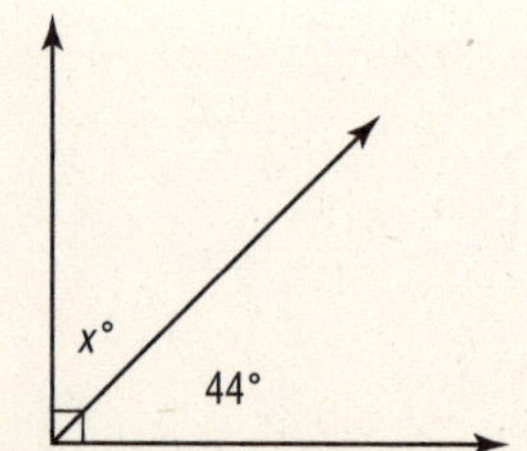

20.
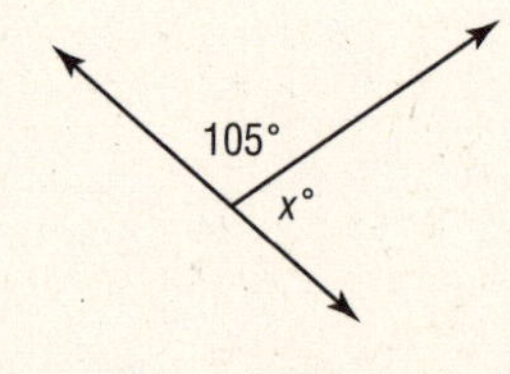

10-2 Skills Practice

Congruent Triangles

For each pair of congruent triangles, name the corresponding parts. Then complete the congruence statement.

1.
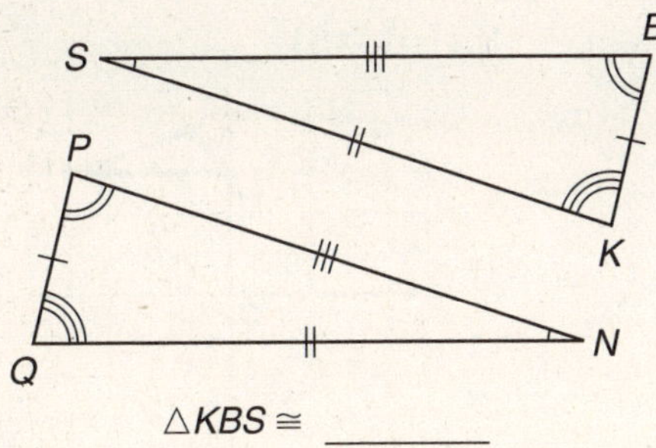

2.
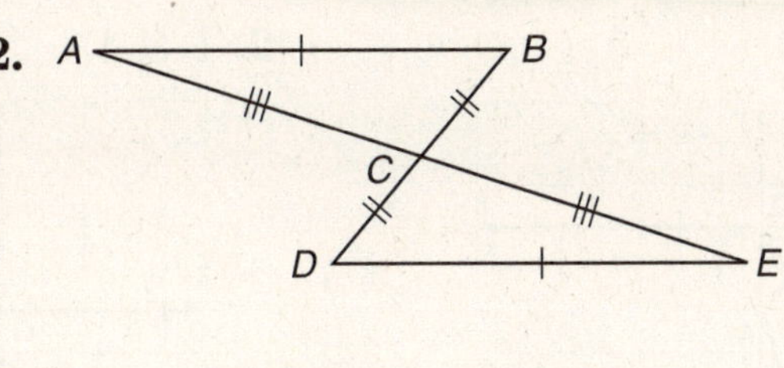

Complete each congruence statement if $\triangle MRU \cong \triangle ACF$.

3. $\angle R \cong$ __?__   4. $\overline{CA} \cong$ __?__   5. $MU \cong$ __?__   6. $\angle A \cong$ __?__

Complete each congruence statement if $\triangle GLE \cong \triangle SPT$.

7. $\overline{EL} \cong$ __?__   8. $\angle S \cong$ __?__   9. $\angle E \cong$ __?__   10. $\overline{PS} \cong$ __?__

Find the value of x for each pair of congruent triangles.

11.
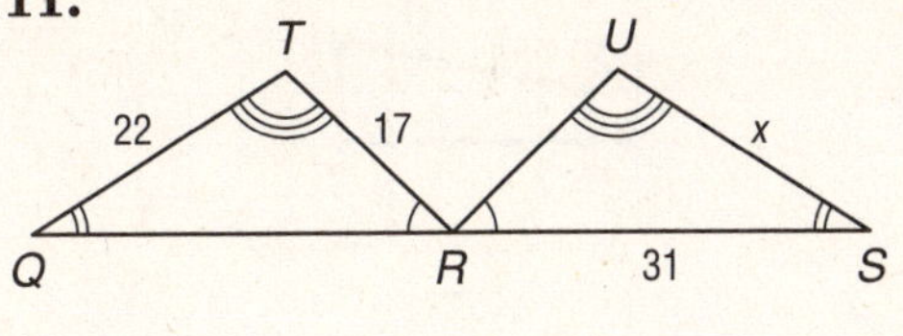

12.
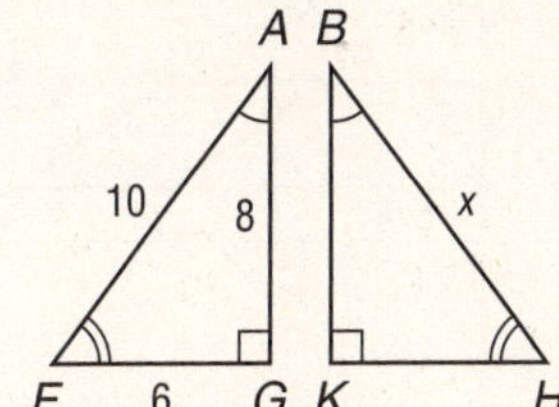

13.
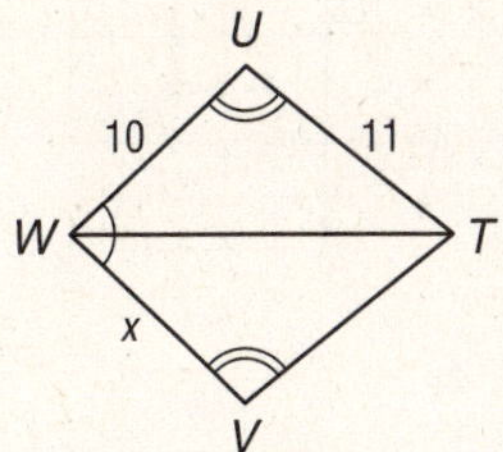

ARCHITECTURE For Exercises 14 and 15, use the diagram of the Eiffel Tower truss at the right and the fact that $\triangle ACB \cong \triangle DFE$.

14. Find the distance between A and B.

15. What is the measure of $\angle B$?

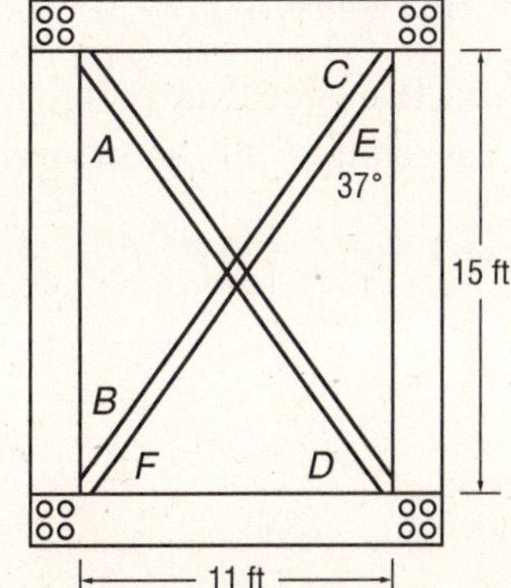

10-3 Skills Practice

Transformations on the Coordinate Plane

Find the coordinates of the vertices of each figure after the given translation. Then graph the translation image.

1. (5, 2)

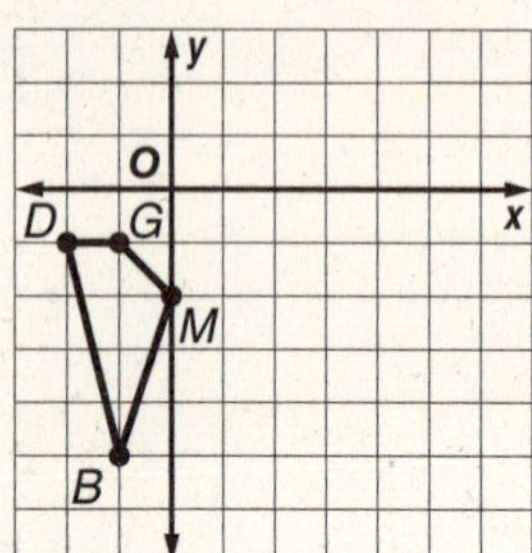

2. (−3, 4)

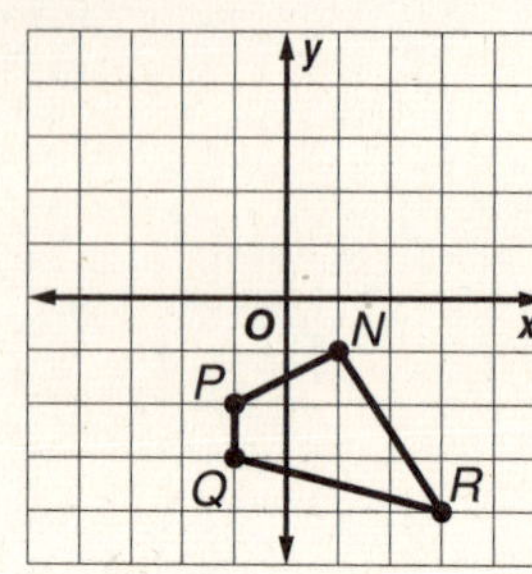

3. (−1, −5)

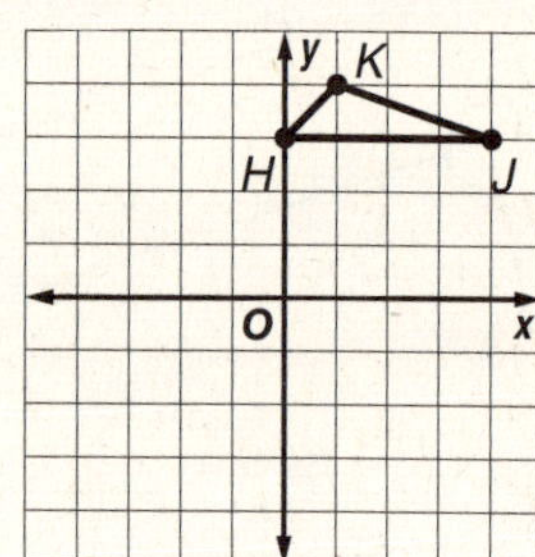

Find the coordinates of the vertices of each figure after a reflection over the given axis. Then graph the reflection image.

4. y-axis

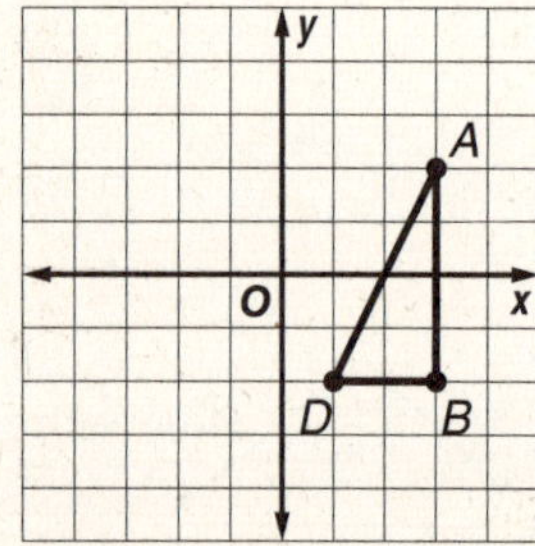

5. x-axis

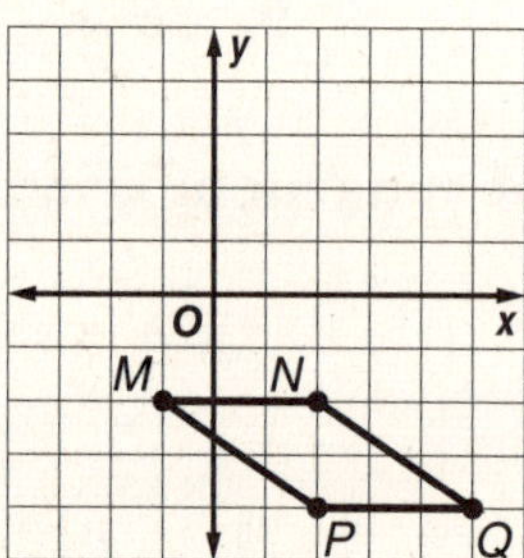

6. x-axis

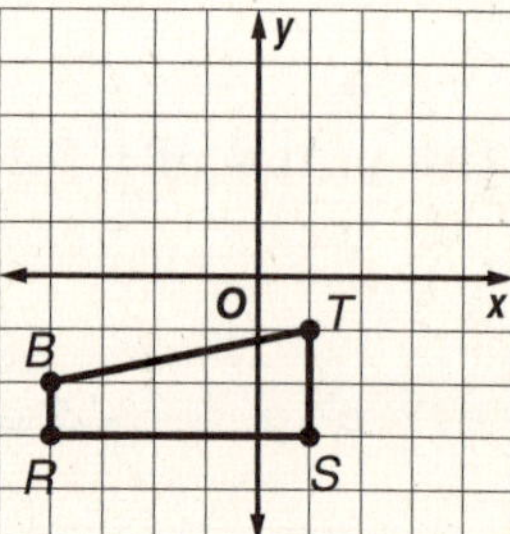

For Exercises 7–9, use the graph shown.

7. Graph the image of the figure after a rotation of 180°.

8. Find the coordinates of the vertices of the figure after a rotation of 90° counterclockwise.

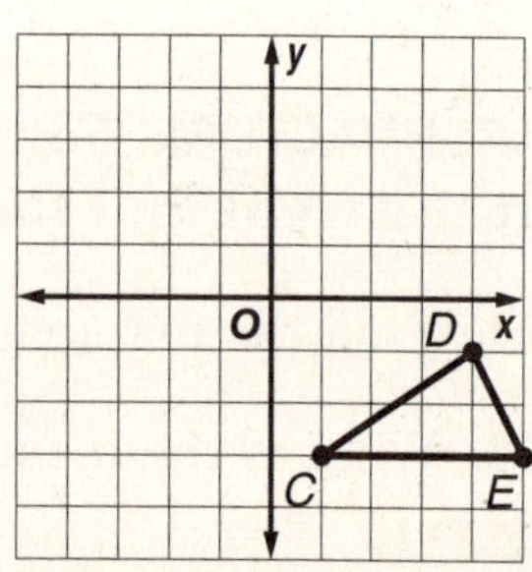

10-4 Skills Practice

Quadrilaterals

ALGEBRA Find the value of *x*. Then find the missing angle measures.

1. 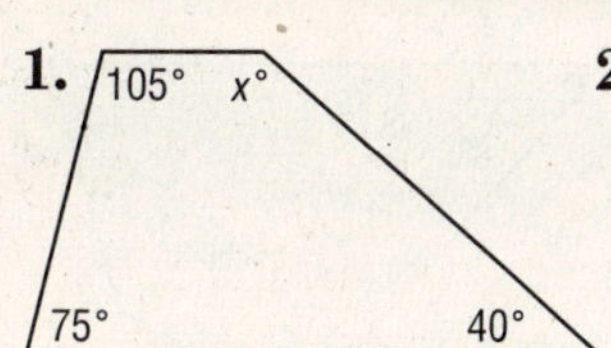	**2.** 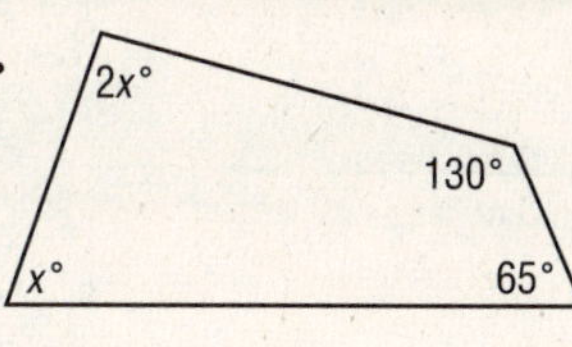	**3.** 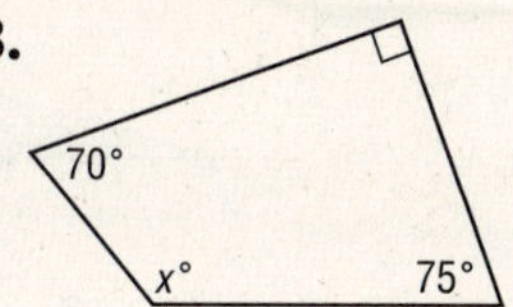	**4.** 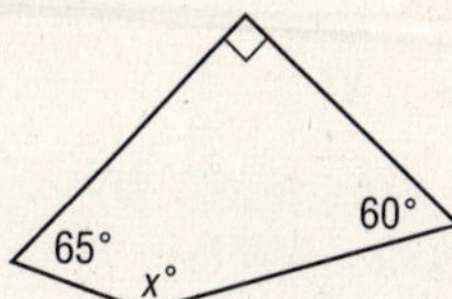
5. 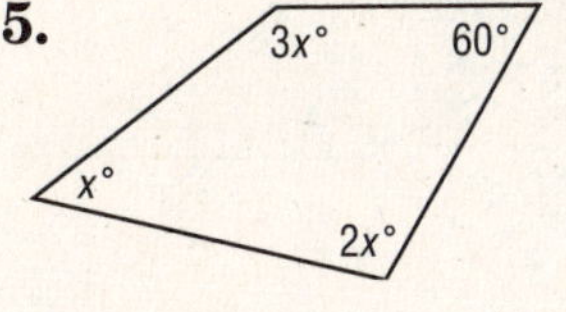	**6.** 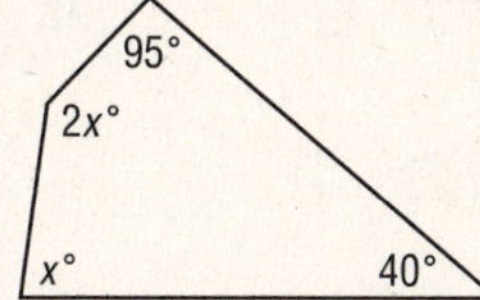	**7.** 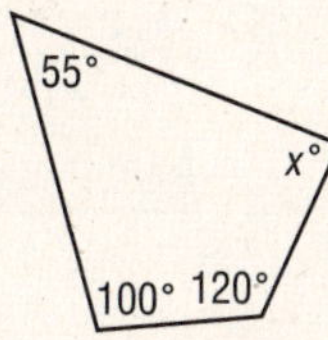	**8.** 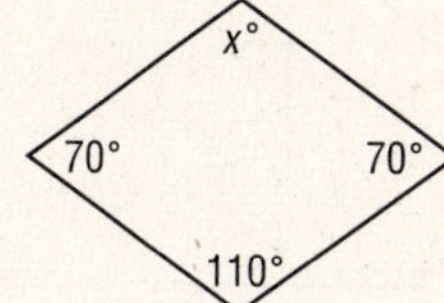
9. 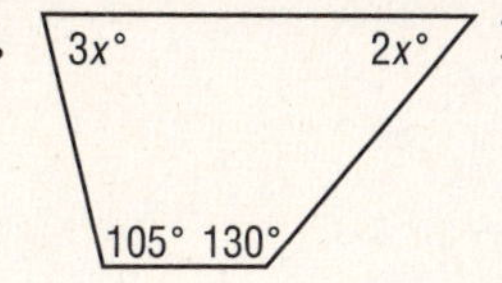	**10.** 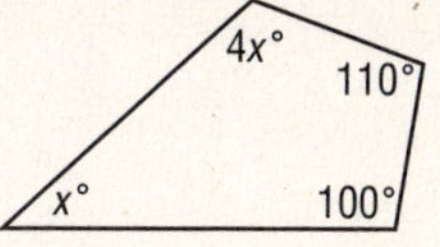	**11.** 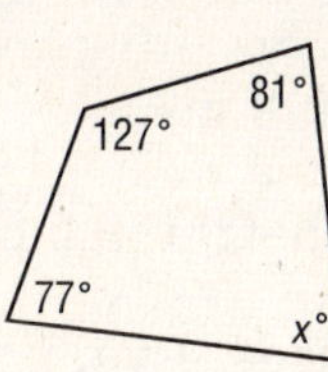	**12.** 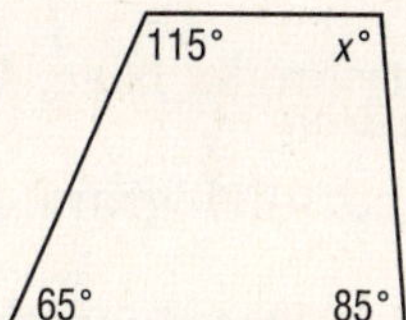

Classify each quadrilateral using the name that *best* describes it.

13. 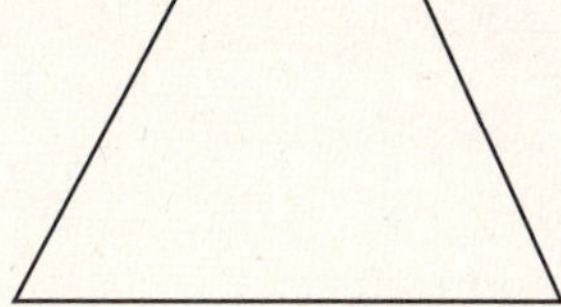	**14.** 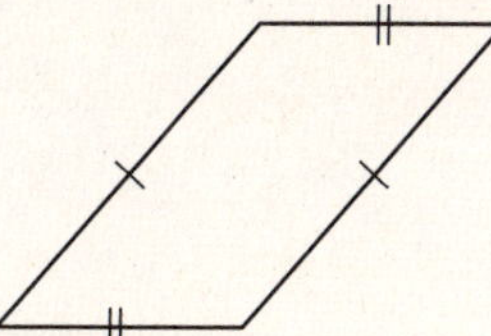	**15.** 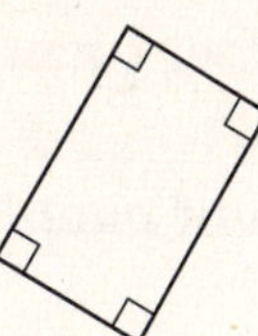
16. 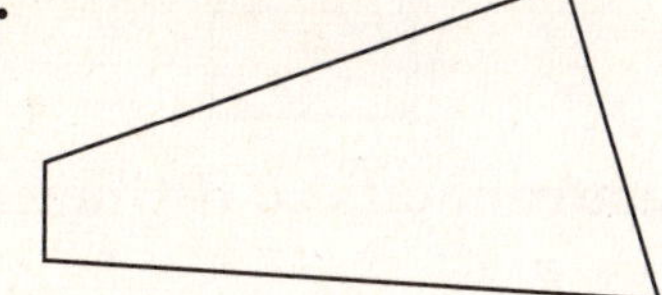	**17.** 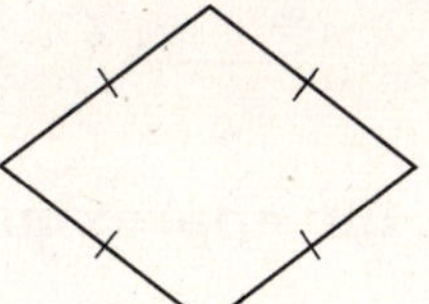	**18.** 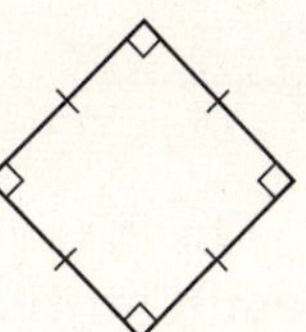

Tell whether each statement is *sometimes*, *always*, or *never* true.

19. A rhombus is a square.

20. A square is a parallelogram.

21. A parallelogram is a square.

10-5 Skills Practice

Area: Parallelograms, Triangles, and Trapezoids

Find the area of each figure.

1. 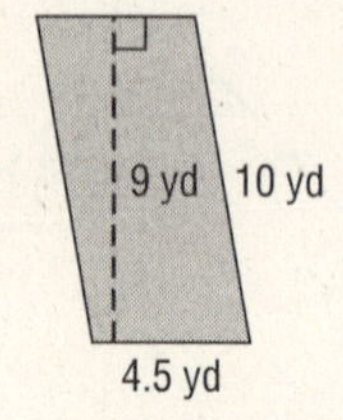

2. 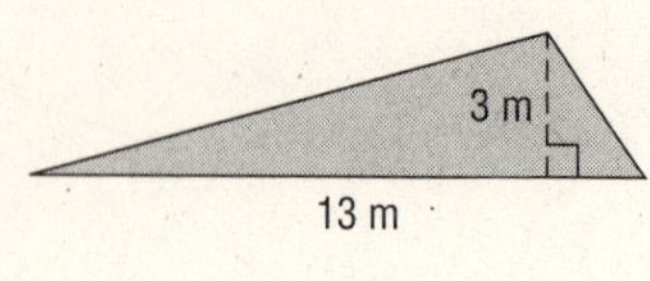

3. 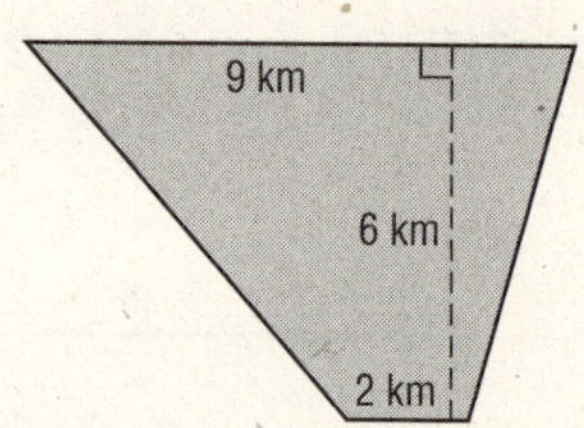

4. 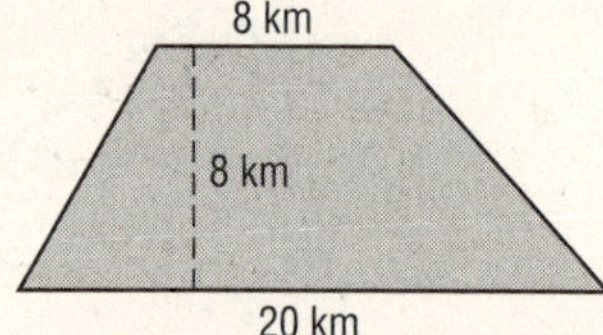

5. 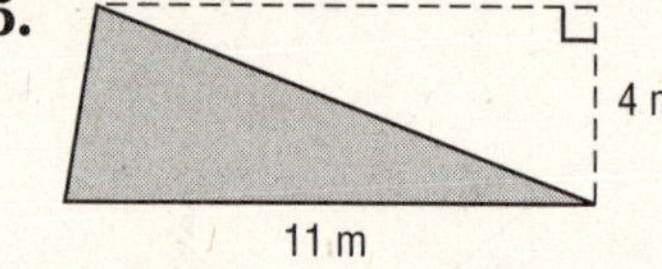

6. 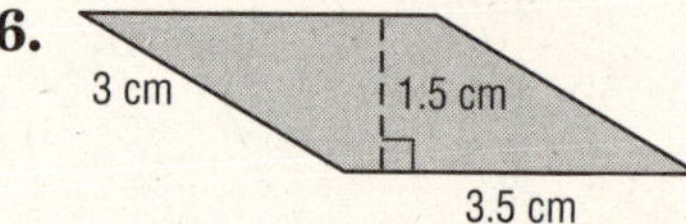

Find the area of each figure described.

7. triangle: base, 11 m; height, 3 m

8. parallelogram: base, 8 cm; height, 9.5 cm

9. trapezoid: height, 12 yd; bases, 4 yd, 7 yd

10. parallelogram: base, 6.5 ft; height, 12 ft

11. trapezoid: height, 10 m; bases, 3 m, 6 m

12. triangle: base, 7 km; height, 5 km

Find the area of each figure.

13. 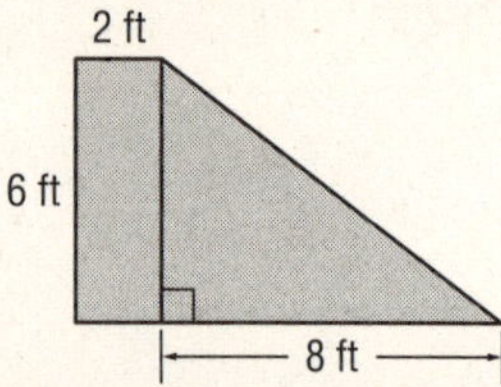

14. 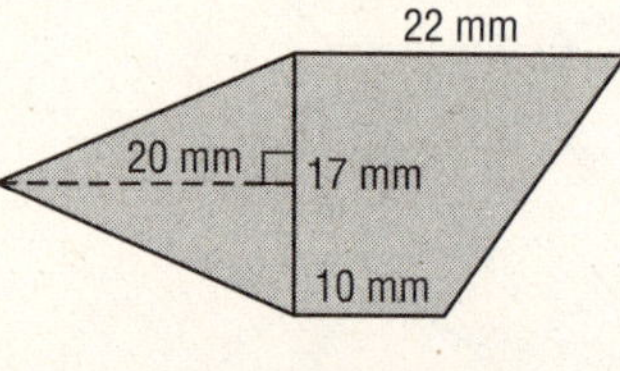

15. 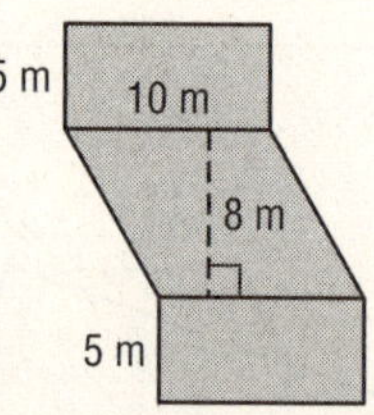

GEOGRAPHY For Exercises 16–18, use the approximate measurements to estimate the area of each state.

16. Alabama

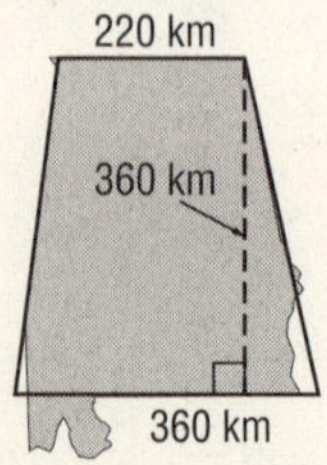

17. Florida

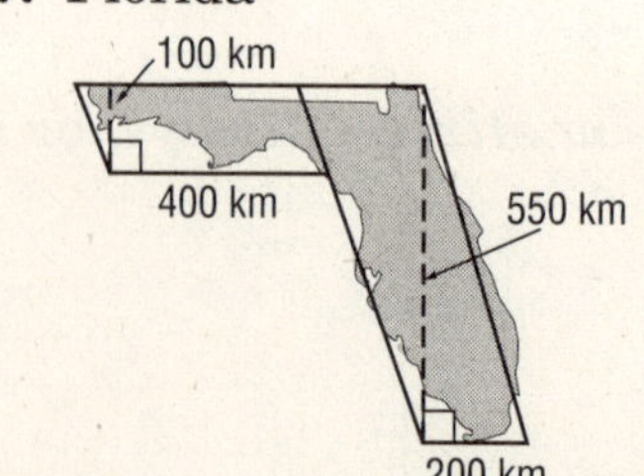

18. Nevada

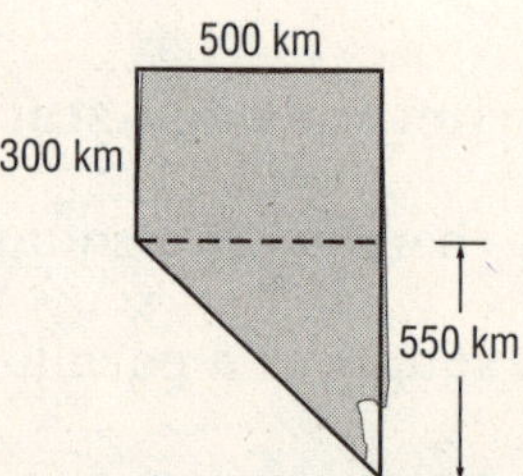

10-6 **Skills Practice**

Polygons

Classify each polygon. Then determine whether it appears to be *regular* or *not regular*.

1. 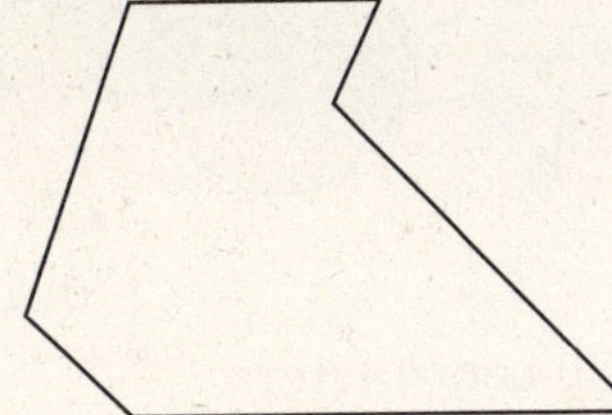

2. 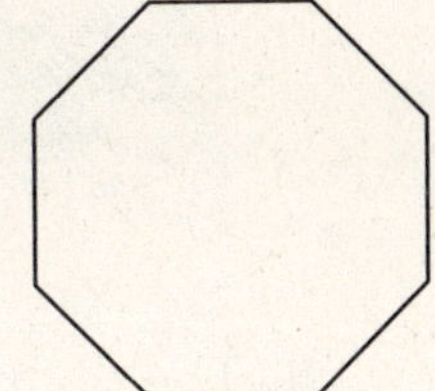

3. 

4. 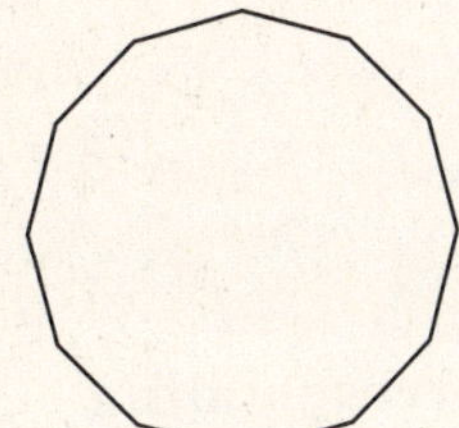

5. 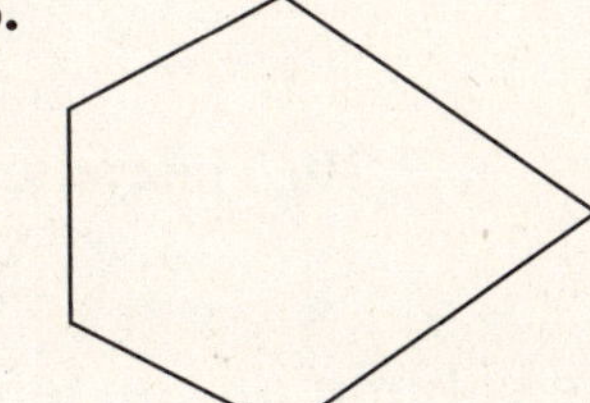

6. 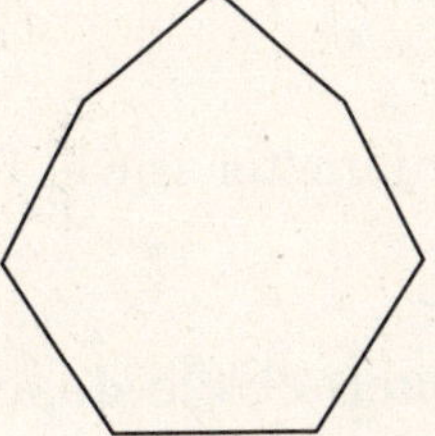

Find the sum of the measures of the interior angles of each polygon.

7. pentagon

8. 20–gon

9. nonagon

10. decagon

Find the measure of an interior angle of each polygon.

11. regular hexagon

12. regular heptagon

13. regular quadrilateral

14. regular octagon

15. regular pentagon

16. regular 100–gon

TESSELLATIONS For Exercises 17 and 18, identify the polygons used to create each tessellation.

17. 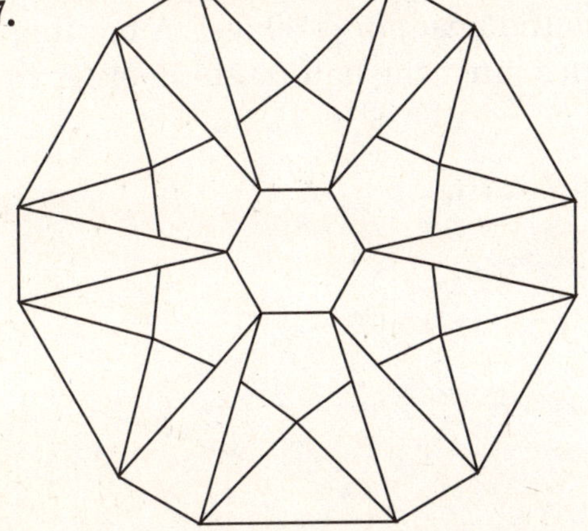

18. 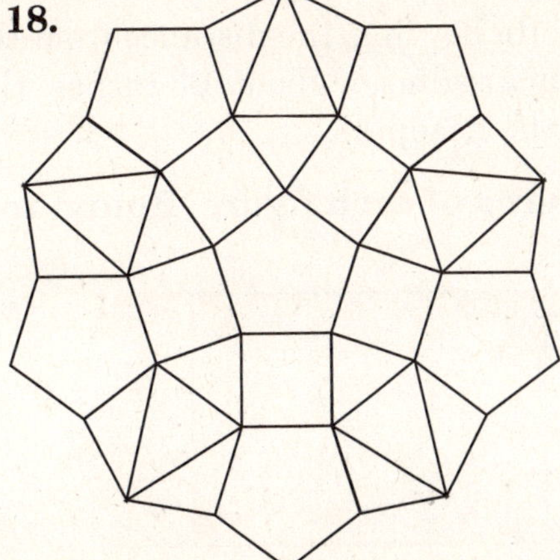

Glencoe Pre-Algebra

10-7 Skills Practice

Circumference and Area: Circles

Find the circumference and area of each circle. Round to the nearest tenth.

1. 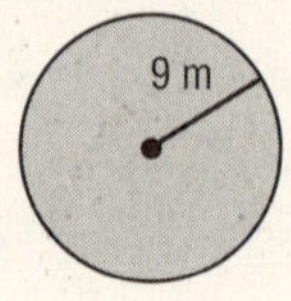

2. 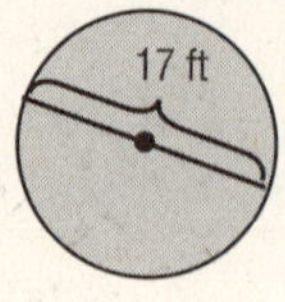

3. 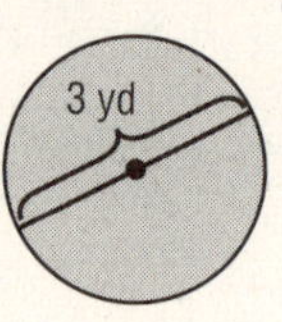

4. 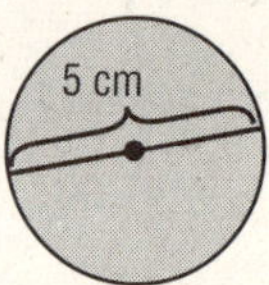

5. The radius is 7 kilometers.

6. The diameter is 20 centimeters.

7. The diameter is 8.5 meters.

8. The radius is 11 yards.

9. The diameter is $6\frac{2}{5}$ feet.

10. The radius is 25 inches.

Match each circle described in the column on the left with its corresponding measurement in the column on the right.

11. diameter: 6 units

   a. area: 19.6 units^2

12. radius: 9 units

   b. circumference: 40.8 units

13. diameter: 13 units

   c. area: 28.3 units^2

14. radius: 2.5 units

   d. circumference: 56.5 units

15. SPORTS A basketball goal is 18 inches in diameter. A basketball has a diameter of about 9.6 inches. What is the difference in area between the goal and the center cross-section of a basketball?

16. CULTURE The Navajo and Pueblo Indians create large, circular sand paintings as part of traditional healing ceremonies. How much more area does a sand painting with a 20-foot diameter have compared with one with a 5-foot diameter?

17. SPORT In bowling, the distance from the foul line to the headpin is 60 feet. A bowling ball has a radius of about 4.3 inches. How many times must the ball rotate in order to strike the headpin?

Find the area of each figure. Round to the nearest tenth.

18. 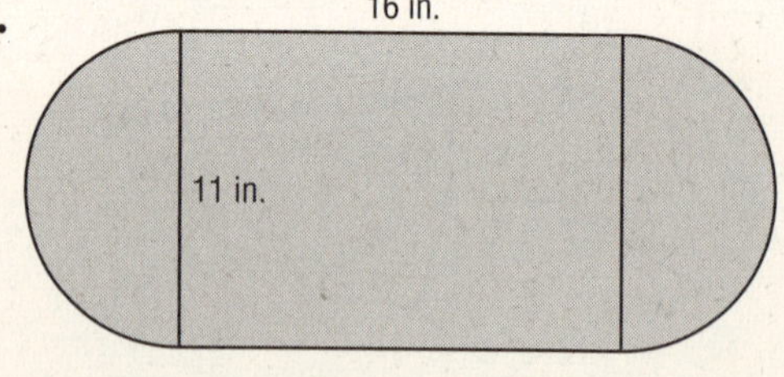

19. 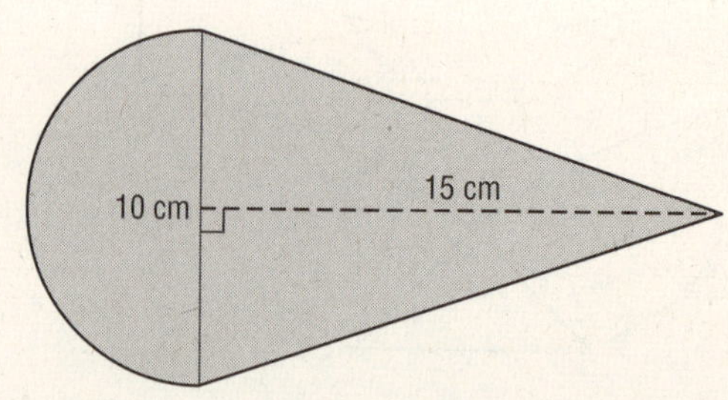

10-8 Skills Practice

Area: Irregular Figures

Find the area of each figure. Round to the nearest tenth.

1. 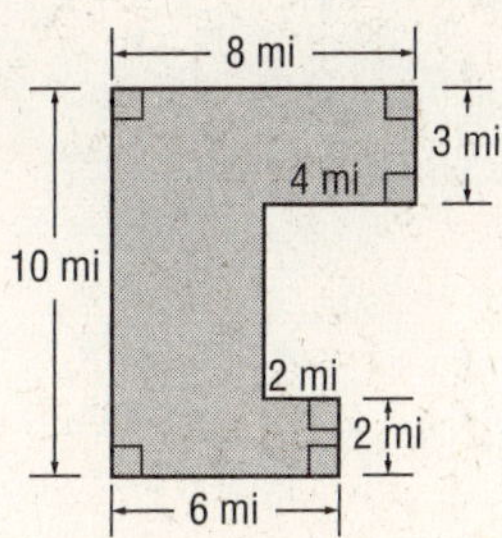

2. 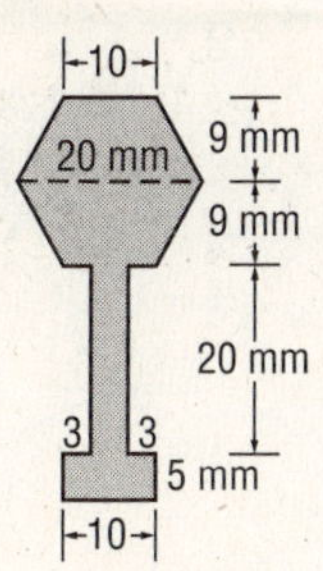

3. 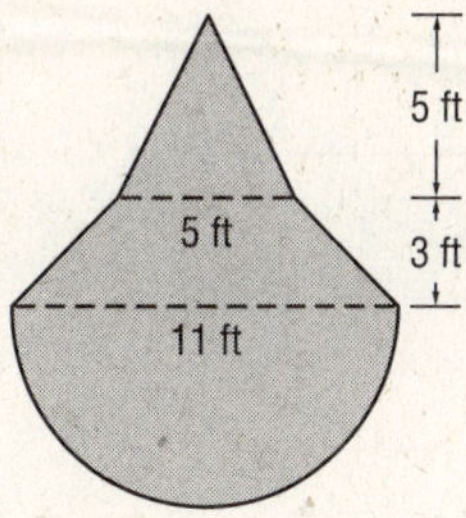

4. 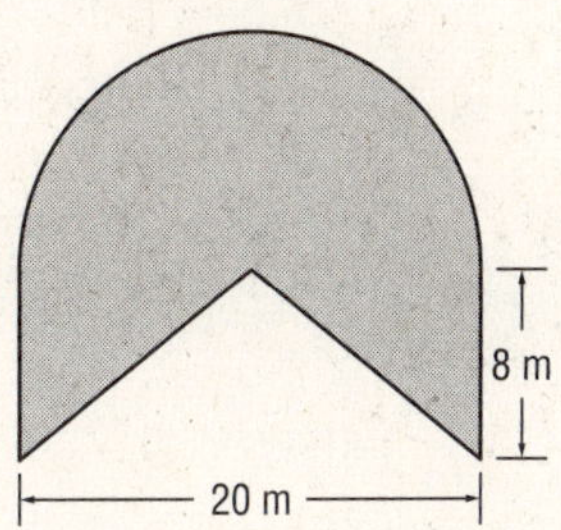

5. 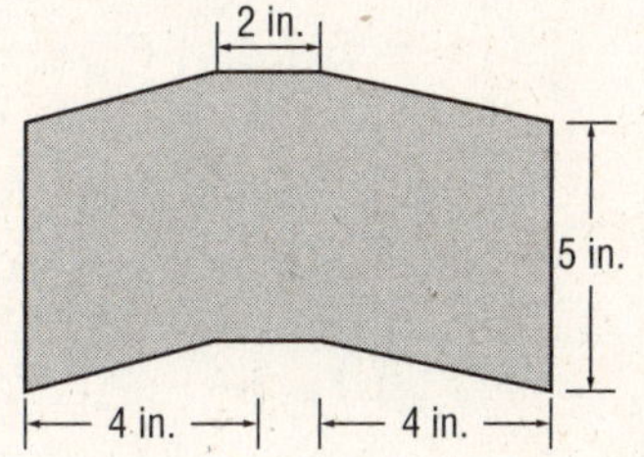

6. 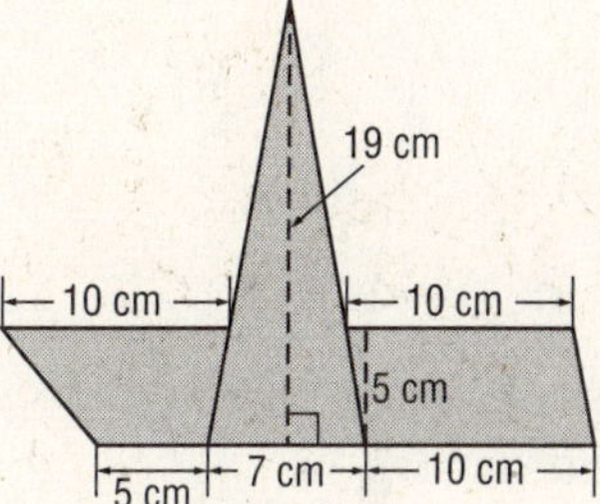

7. 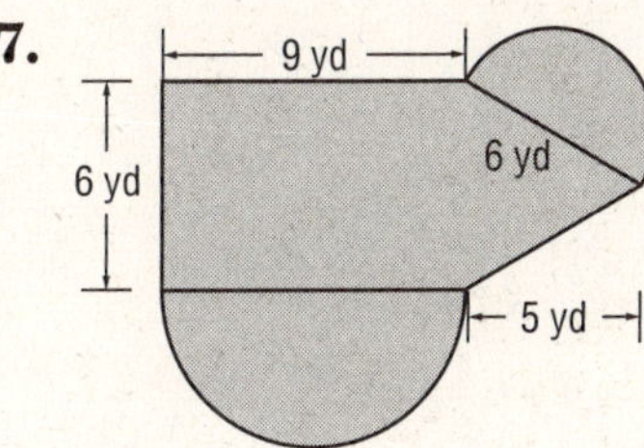

8. 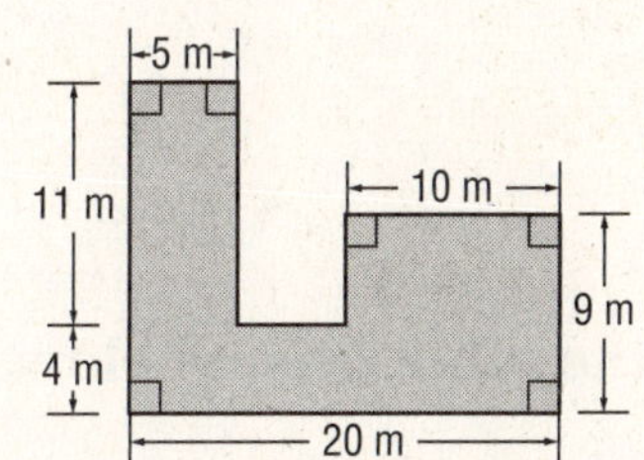

9. 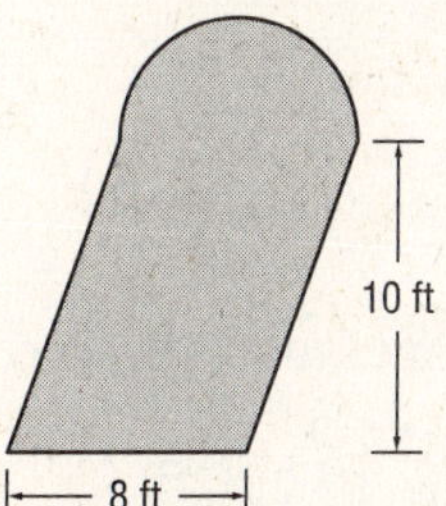

10. What is the area of a figure formed using a square with sides of 12 kilometers and three circles with diameters of 12 kilometers each?

Find the area of each shaded area. Round to the nearest tenth, if necessary. (*Hint:* Find the total area and subtract the non-shaded area.)

11. 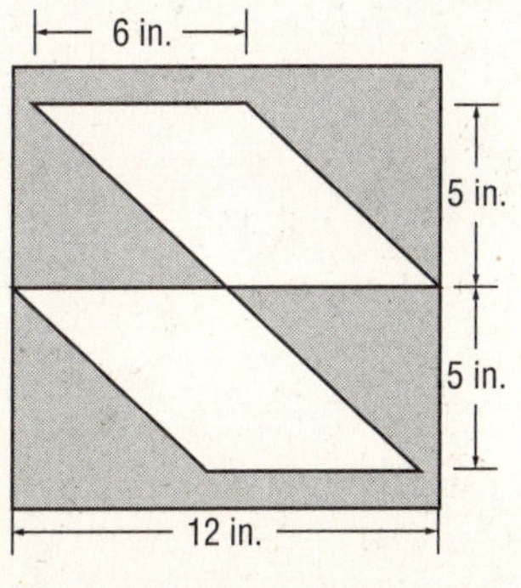

12. 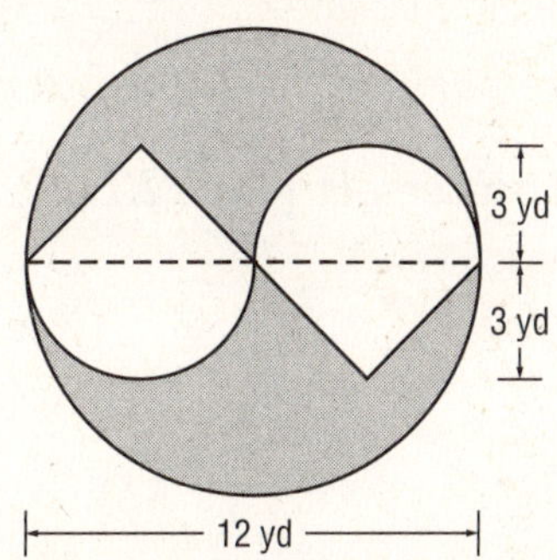

13. 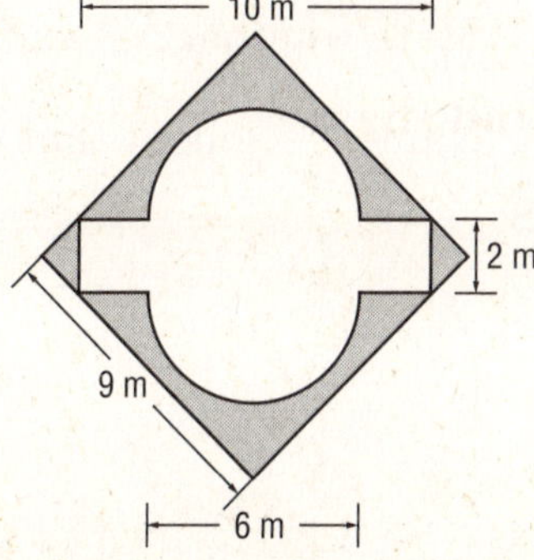

11-1 **Skills Practice**

Three-Dimensional Figures

Identify each solid. Name the bases, faces, edges, and vertices.

1.

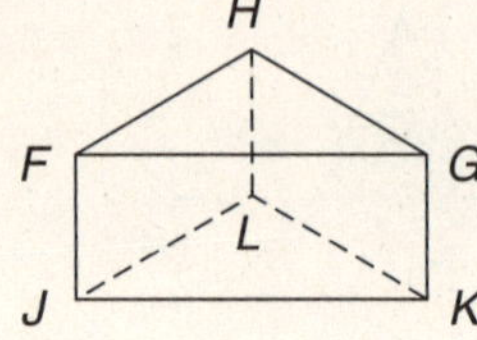

2.

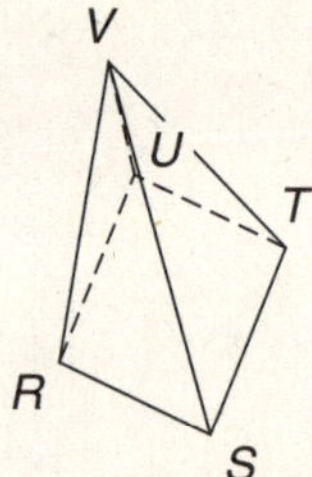

3.

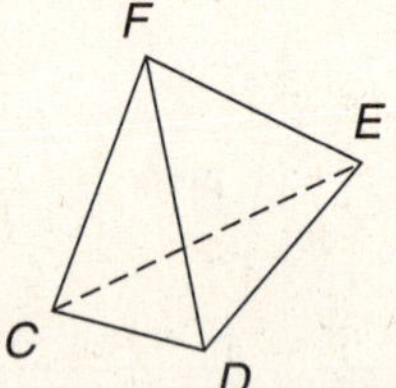

4.

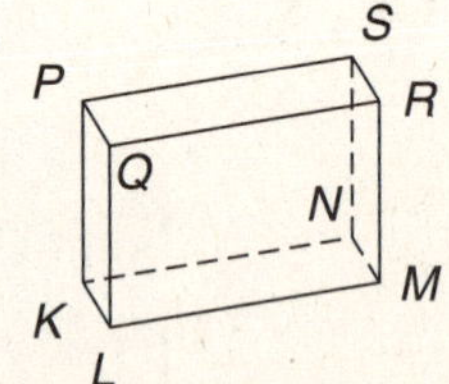

For Exercises 5–8, use the rectangular prism below.

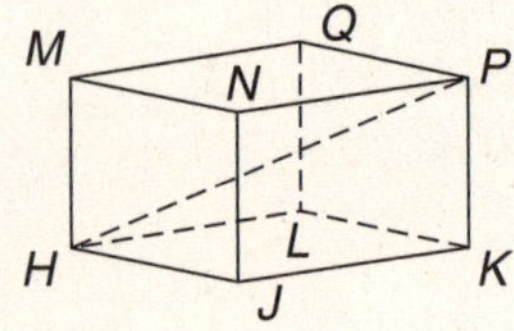

5. Identify a diagonal.

6. Name four segments skew to $\overline{LQ}$.

7. State whether $\overline{NP}$ and $\overline{HM}$ are *parallel*, *skew*, or *intersecting*.

8. Name a segment that does *not* intersect plane *KLQP*.

11-2 Skills Practice

Volume: Prisms and Cylinders

Find the volume of each solid shown or described. If necessary, round to the nearest tenth.

1. 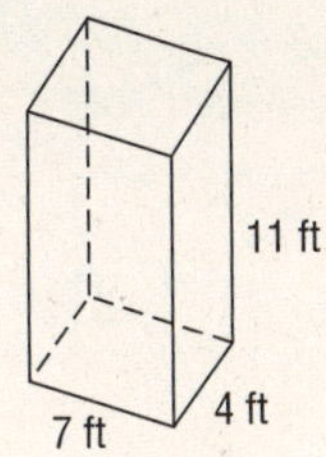

2. 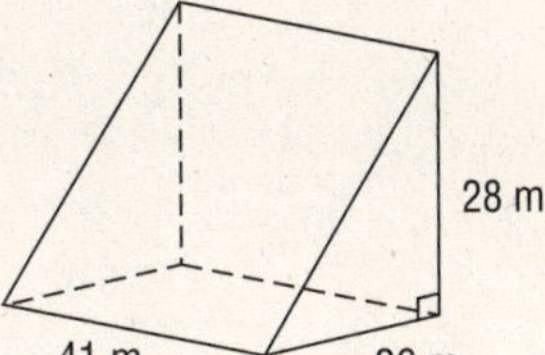

3. 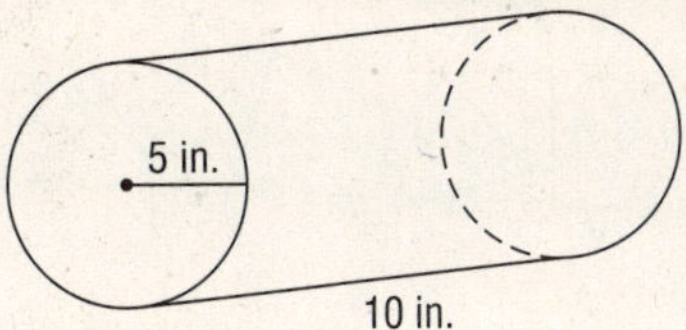

4. 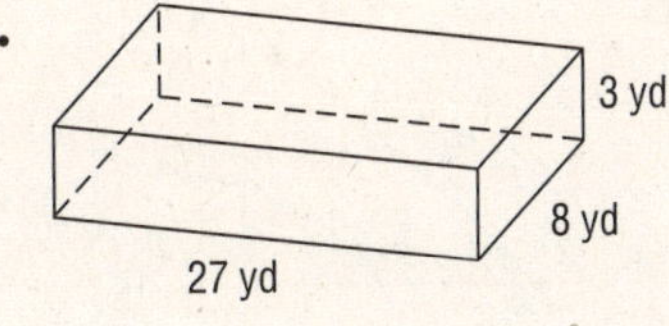

5. 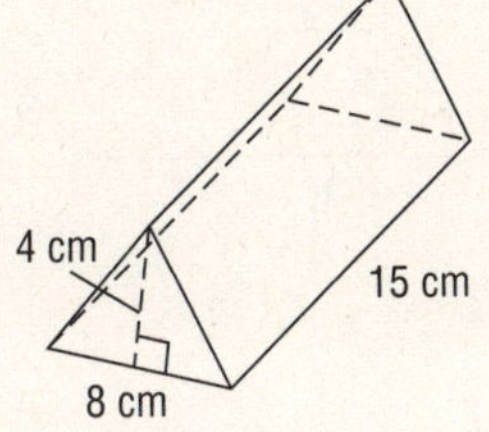

6. 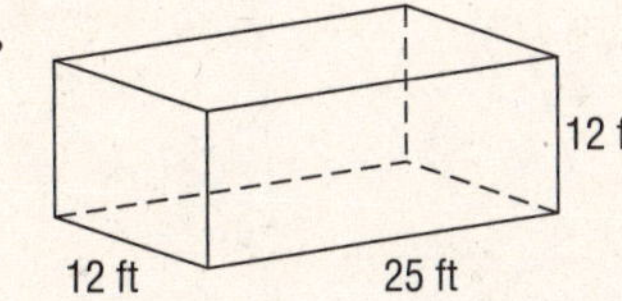

7. 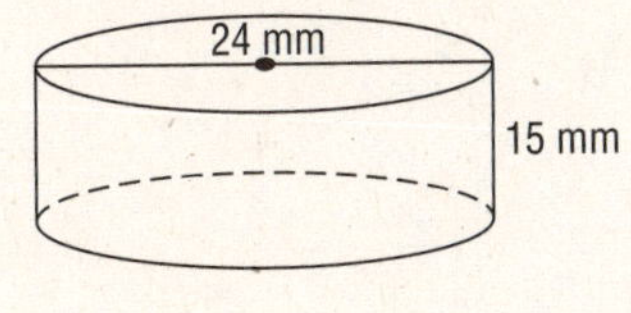

8. 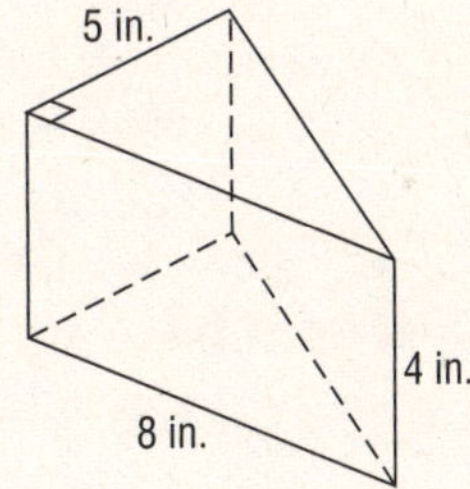

9. 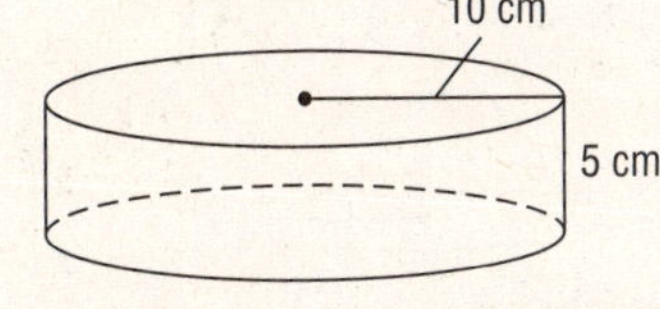

10. rectangular prism: length 18 ft, width 9 ft, height 1 ft

11. triangular prism: base of triangle 22 yd, altitude of triangle 14 yd, height of prism 30 yd

12. Find the height of a cylinder with a radius of 12 inches and a volume of 3754.8 in^3. Round to the nearest tenth.

11-3 Skills Practice

Volume: Pyramids and Cones

Find the volume of each solid. If necessary, round to the nearest tenth.

1. 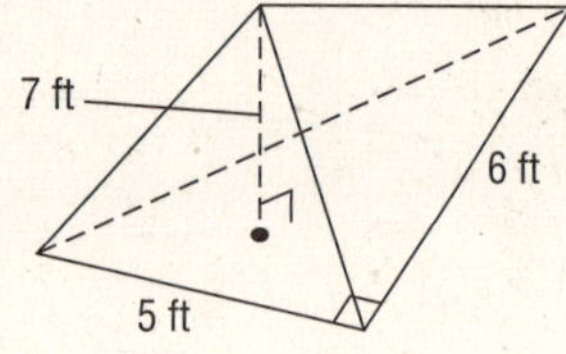

2. 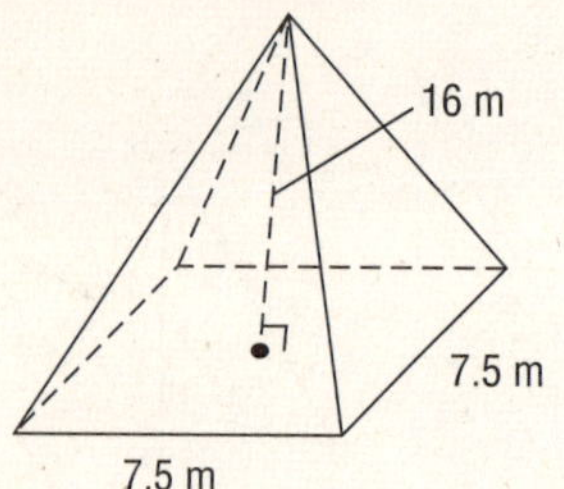

3. 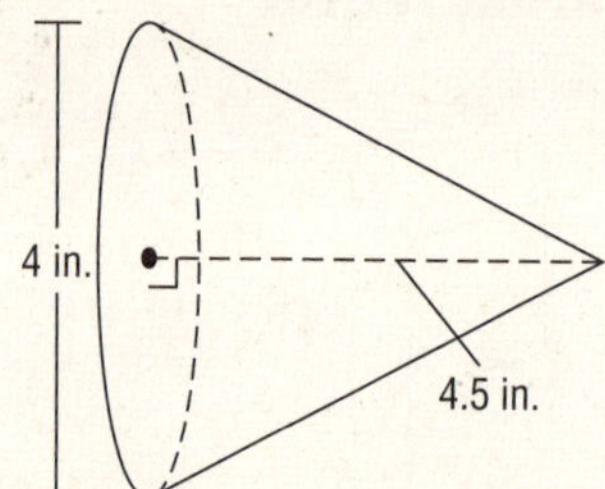

4. 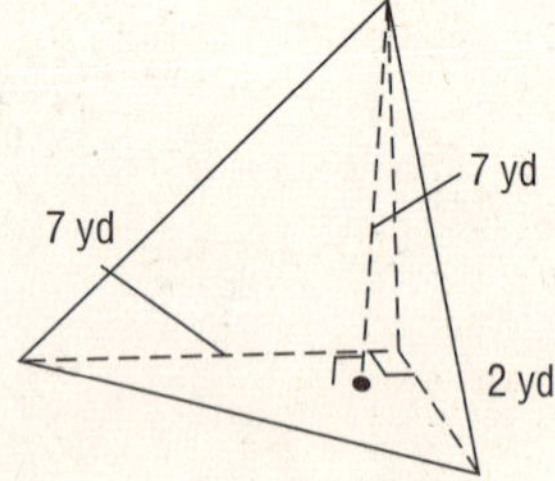

5. 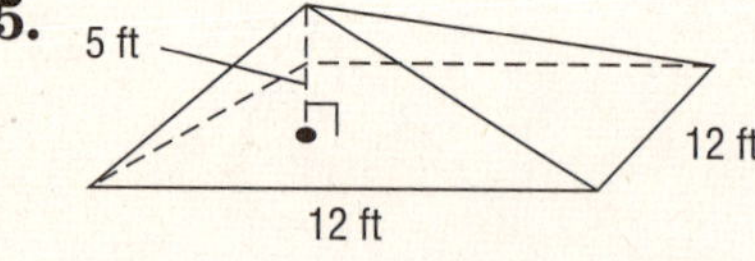

6. 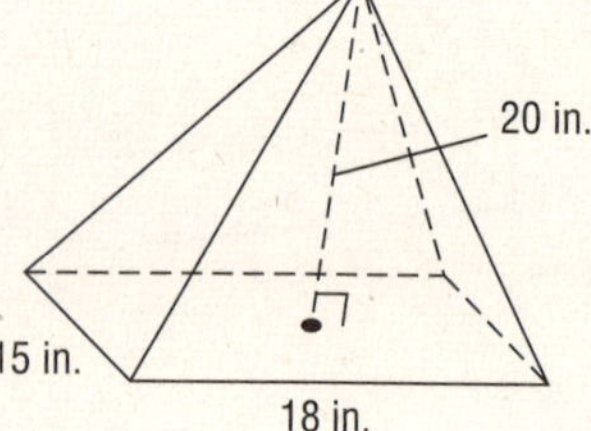

7. 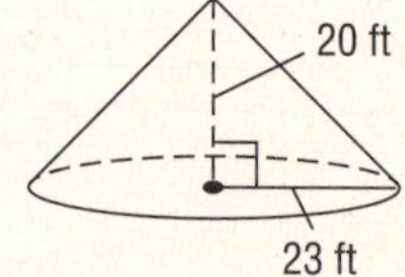

8. 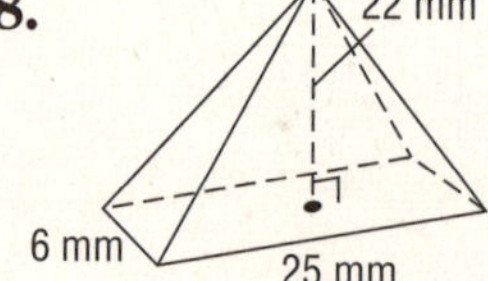

9. 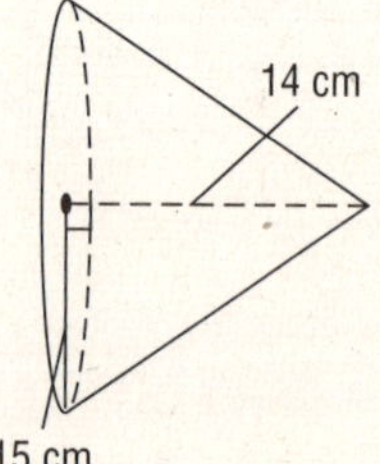

10. rectangular pyramid: length 7 ft, width 2.5 ft, height 8 ft

11. cone: radius 20 cm, height 30 cm

12. Find the height of a cone with a radius of 4 inches and a volume of 160.8 in^3. Round to the nearest tenth

11-4 Skills Practice

Surface Area: Prisms and Cylinders

Find the surface area of each solid shown or described. If necessary, round to the nearest tenth.

1. 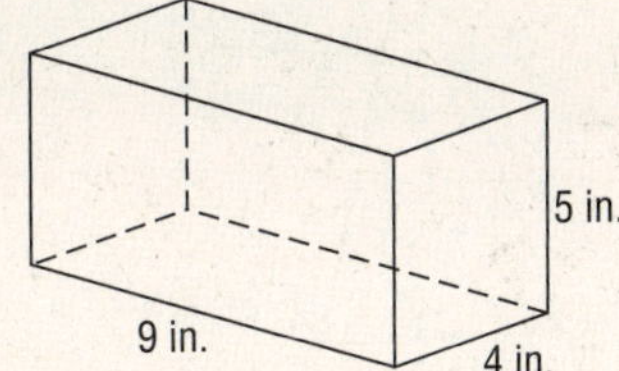

2. 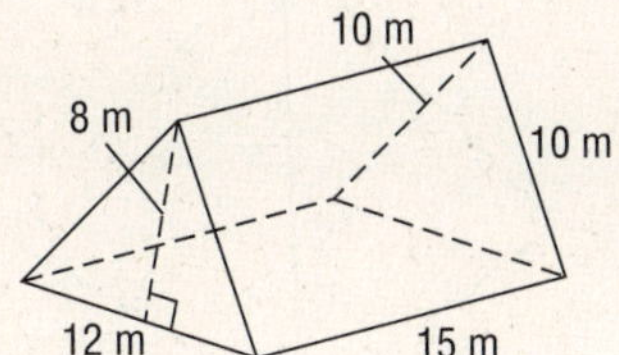

3. 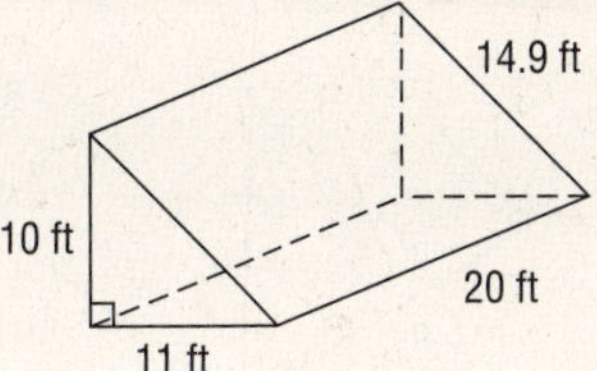

4. 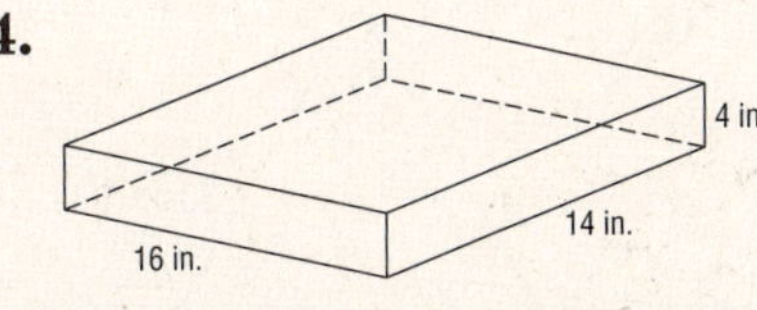

5. 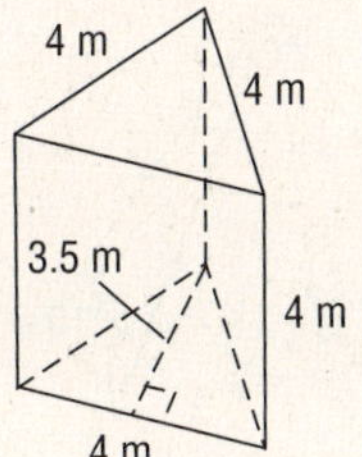

6. 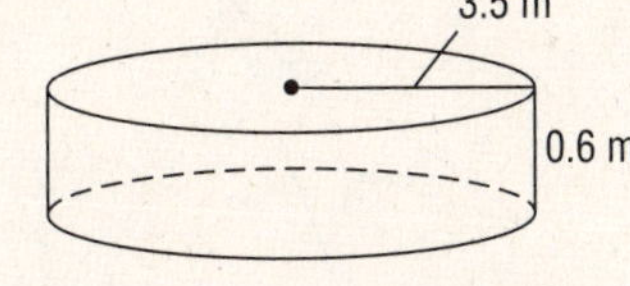

7. 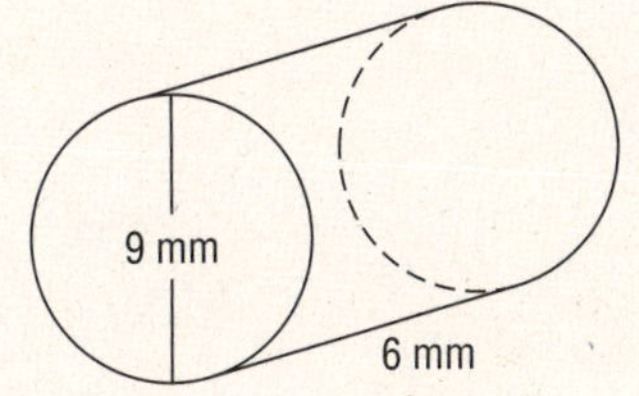

8. 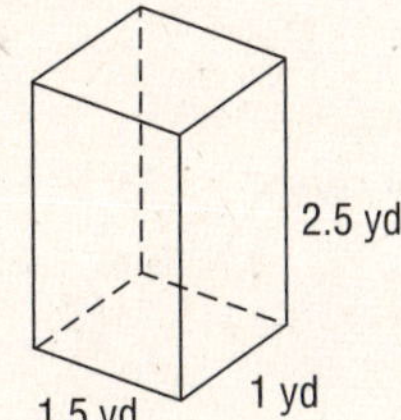

9. 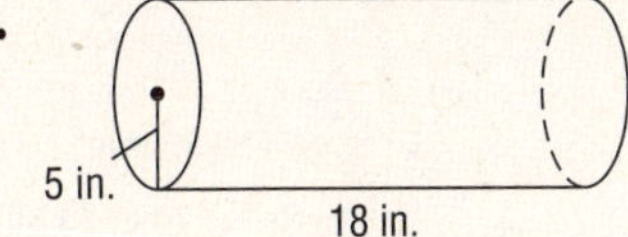

10. rectangular prism: length 17 yd, width 4.5 yd, height 3 yd

11. cylinder: radius 16 ft, height 42 ft

12. cylinder: diameter 20.2 cm, height 43 cm

11-5 Skills Practice

Surface Area: Pyramids and Cones

Find the surface area of each solid. If necessary, round to the nearest tenth.

1.
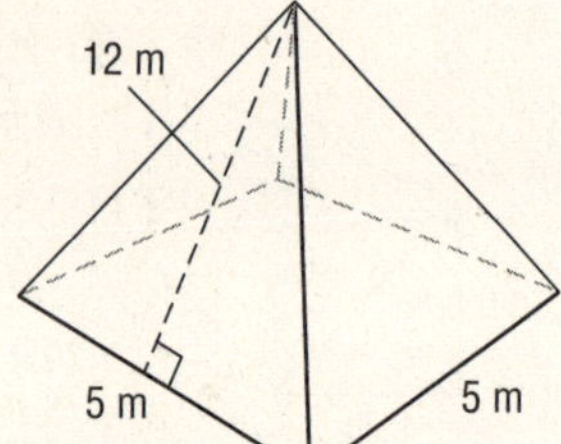

2.
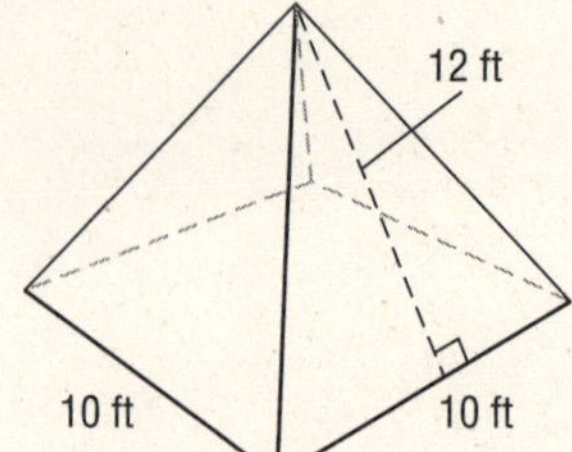

3.
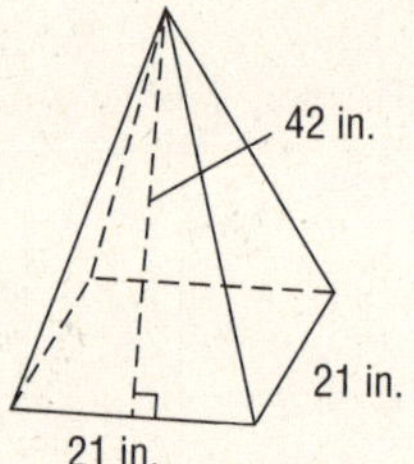

4.
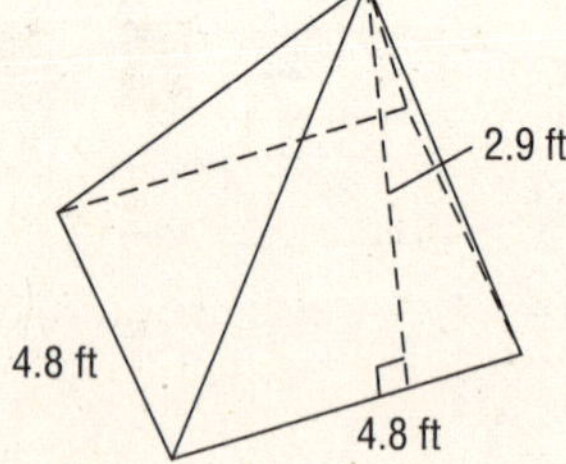

5.
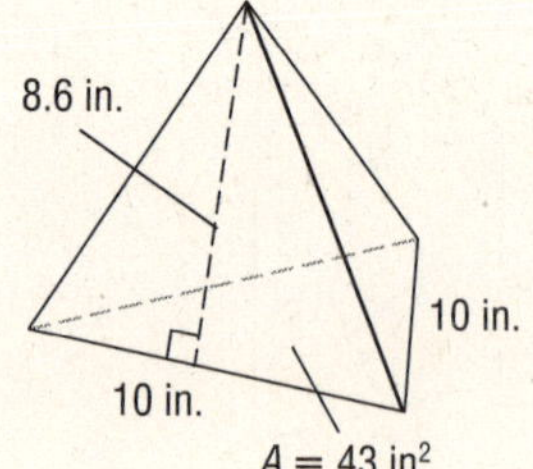

6.
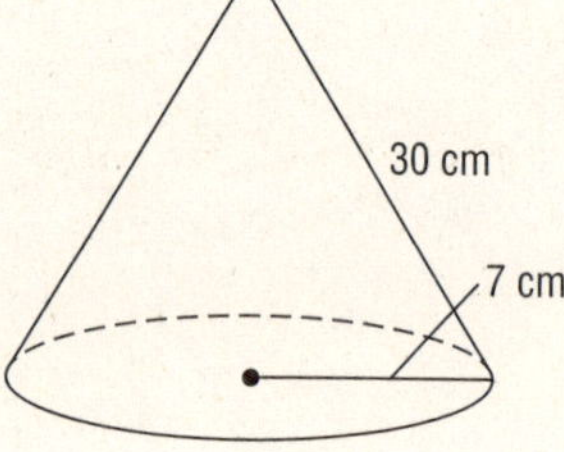

7.
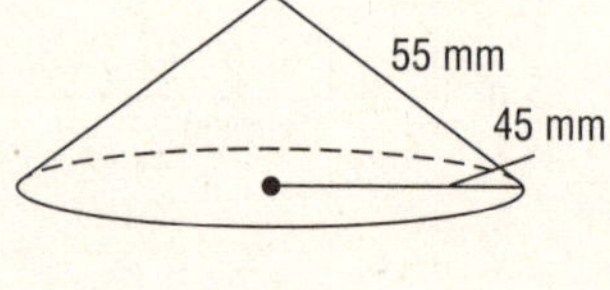

8.
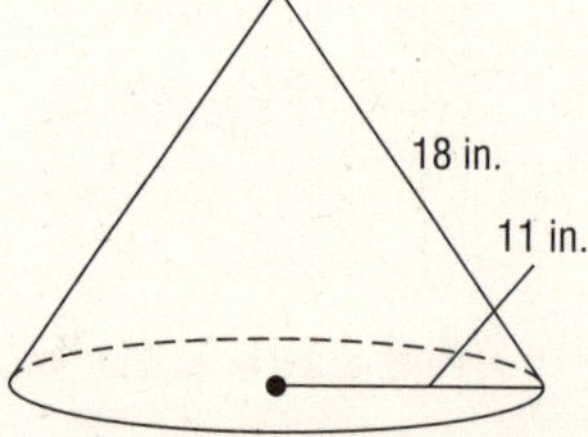

9.
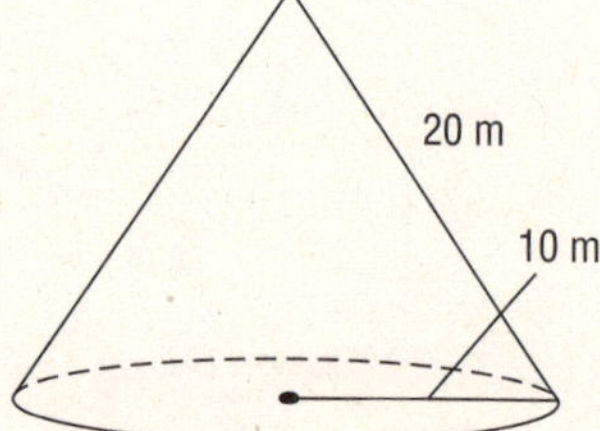

10. square pyramid: base side length 6.3 m, slant height 4 m

11. cone: diameter 16 yd, slant height 10 yd

12. cone: radius 14 cm, slant height 33 cm

Glencoe Pre-Algebra

NAME ______________________________ DATE ____________ PERIOD _____

11-6 Skills Practice

Similar Solids

Determine whether each pair of solids is similar.

1.

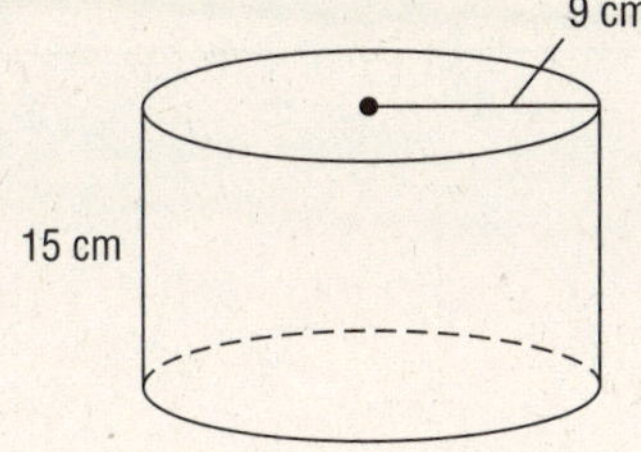

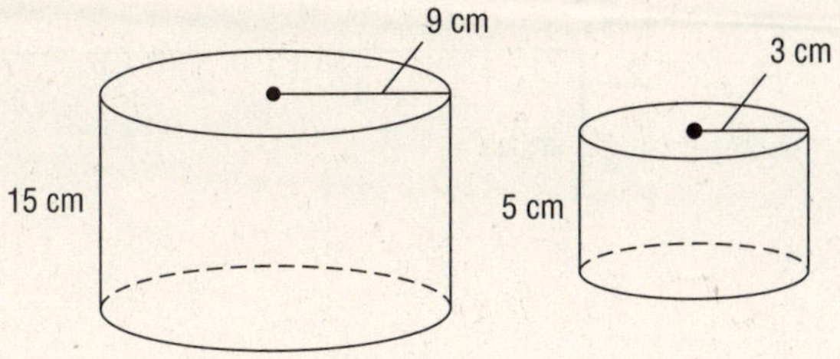

2.

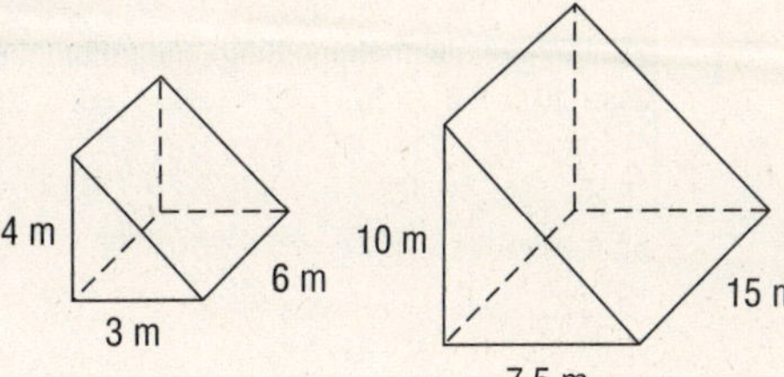

3.

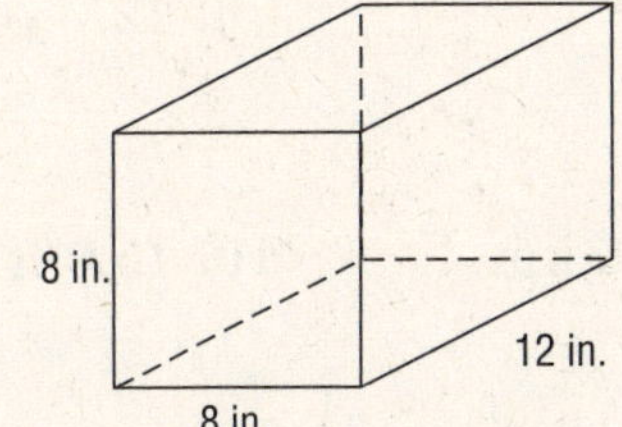

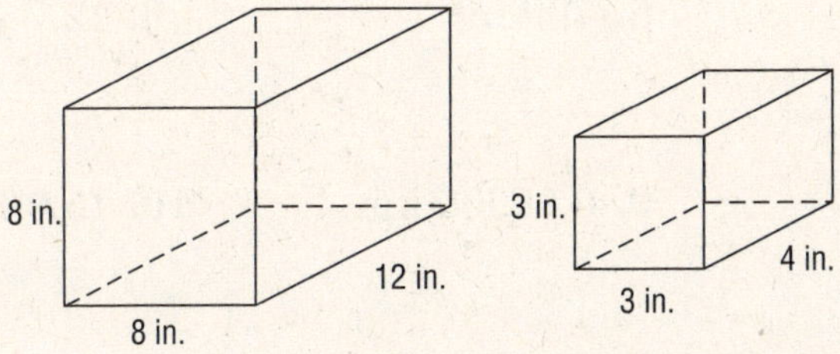

4.

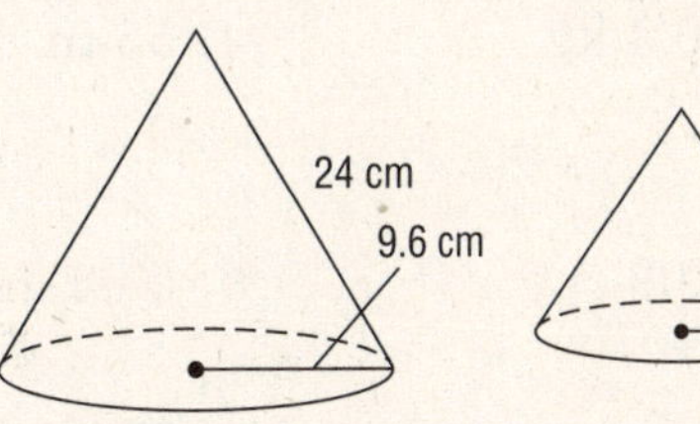

5.

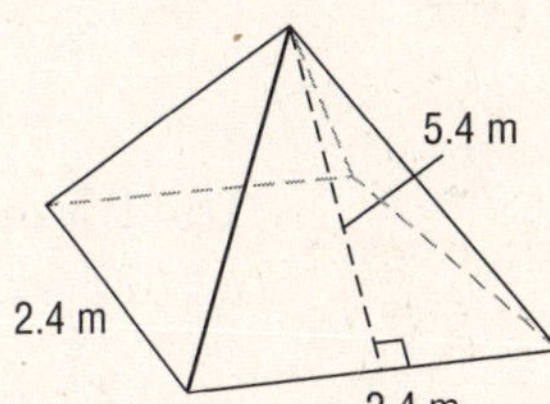

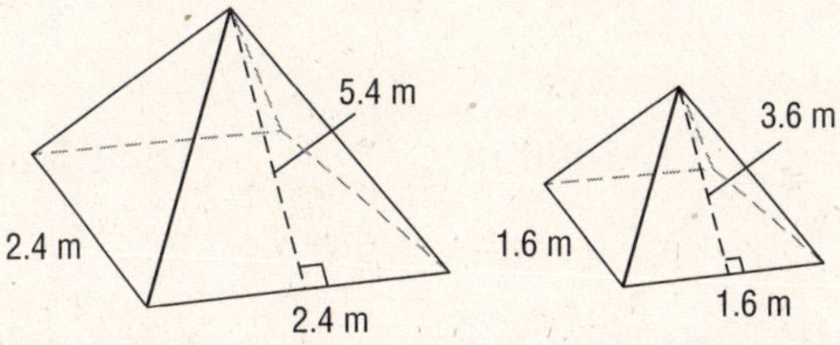

6.

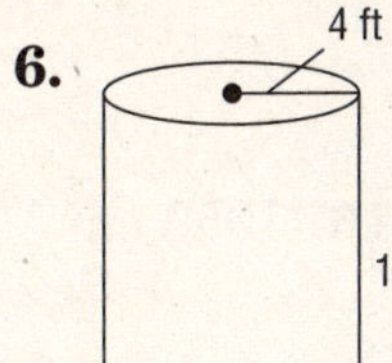

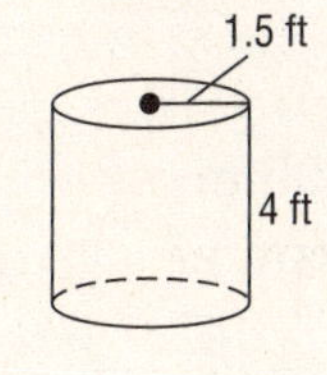

Find the missing measure for each pair of similar solids.

7.

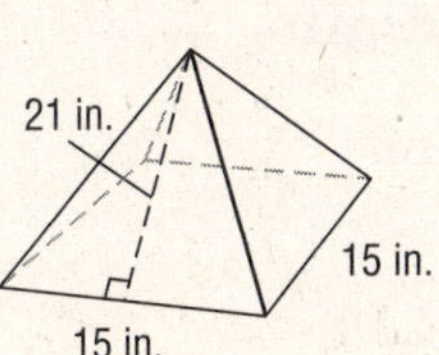

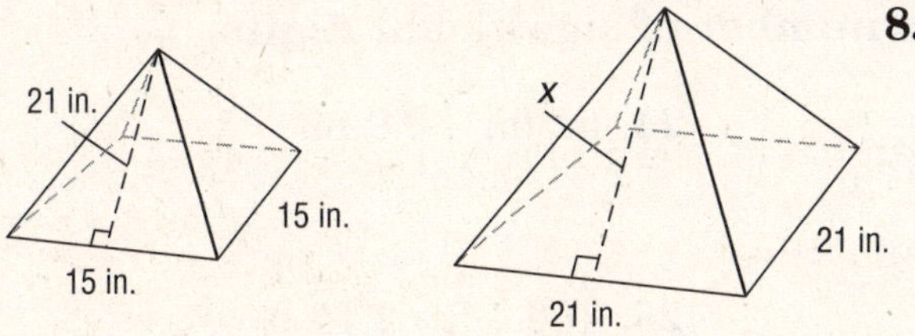

8.

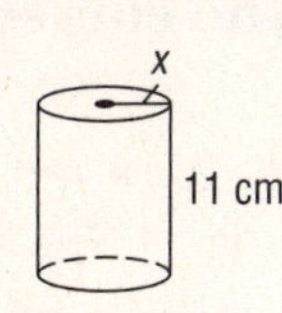

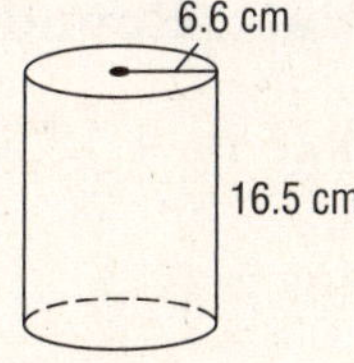

9.

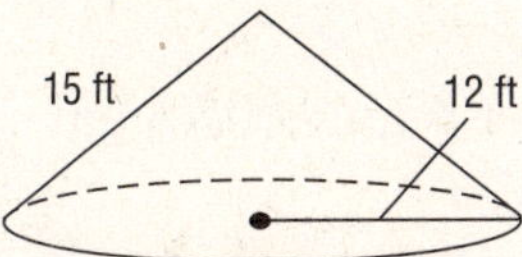

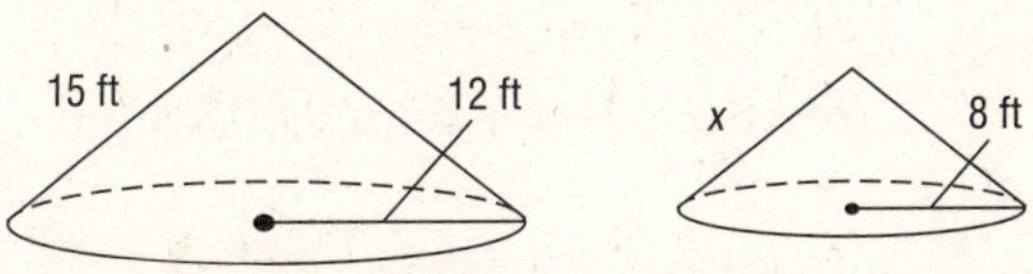

10.

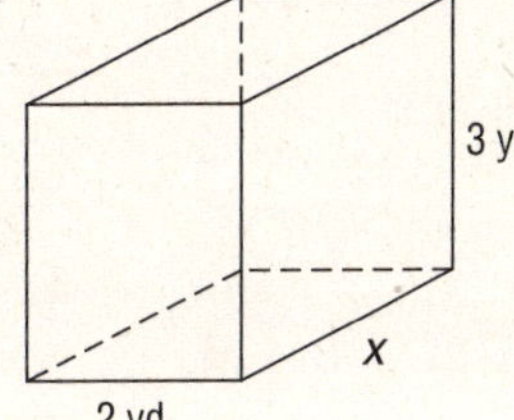

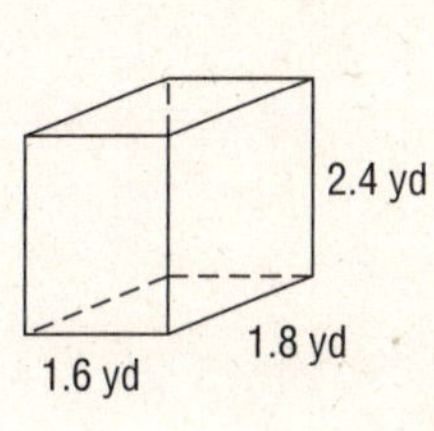

NAME ______________________ DATE ____________ PERIOD _____

11-7 Skills Practice

Precision and Significant Digits

Identify the precision unit of each measuring tool.

1.

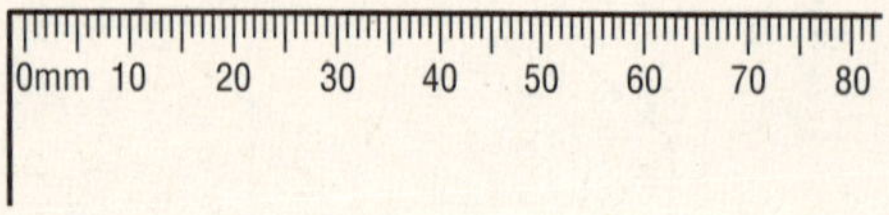

2.

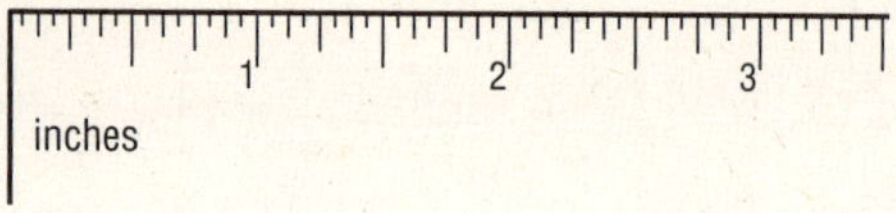

Determine the number of significant digits in each measure.

3. 80.4 kg

4. 55 in.

5. 700 mi

6. 12.7 yd

7. 0.09 cm

8. 5.04 gal

9. 0.08000 km

10. 15.0 ft

11. 6 L

12. 1040 yd

13. 1200 mi

14. 11 mm

15. 1005 m

16. 1050 cm

17. 1500 in.

18. 1.5 mi

19. 0.0015 L

20. 0.01005 km

21. 9.025 ft

22. 8.86 s

Calculate. Round to the correct number of significant digits.

23. 40.25 in. + 12.5 in.

24. 51 mi – 0.9 mi

25. 7.6 cm + 1.04 cm + 2.35 cm

26. 180.4 ft · 2.75 ft

27. 17.3 cm · 8 cm

28. 0.2 gal + 1.0 gal + 0.75 gal

29. 1.2 cL + 11 cL + 3.4 cL

30. 5.030 km · 4.001 km

31. 14.5 yd – 6.875 yd

32. 0.081 km · 4 km

12-1 **Skills Practice**

Stem-and-Leaf Plots

Display each set of data in a stem-and-leaf plot.

1. {7, 2, 3, 11, 20, 21, 17, 15, 15, 14}

2. {8, 2, 14, 27, 7, 2, 16, 13, 29, 16}

3.

Amount of Fresh Fruit Consumed per Person in the United States, 1998	
Fruit	**Pounds Consumed per Person**
Apples	19
Bananas	29
Cantaloupes	11
Grapefruit	6
Grapes	7
Oranges	15
Peaches and nectarines	5
Pears	3
Pineapples	3
Plums and prunes	1
Strawberries	4
Watermelons	15

Source: U.S. Census Bureau

4.

Winning Scores in College Football Bowl Games, 1999	
Game and Winning School	**Points Scored**
Aloha Bowl, Wake Forest	23
Cotton Bowl, Texas	38
Fiesta Bowl, Tennessee	23
Florida Citrus Bowl, Michigan	45
Gator Bowl, Georgia Tech.	35
Holiday Bowl, Kansas St.	24
Liberty Bowl, So. Mississippi	23
Orange Bowl, Florida	31
Outback Bowl, Penn. St.	26
Peach Bowl, Mississippi St.	27
Rose Bowl, Wisconsin	38
Sugar Bowl, Ohio St.	24

Source: *World Almanac*

HUMIDITY   For Exercises 5–7, use the information in the back-to-back stem-and-leaf plot.   **Source:** *The New York Public Library Desk Reference*

5. What is the highest morning relative humidity?

6. What is the lowest afternoon relative humidity?

7. Does relative humidity tend to be higher in the morning or afternoon?

U.S. Average Relative Humidity (percent)

Morning		Afternoon
	5	1 2 3 4 7 9
	6	
8 8 4	7	
9 4 0	8	

8 | 7 = 78%   5 | 3 = 53%

12-2 Skills Practice

Measures of Variation

Find the range and interquartile range for each set of data.

1. {7, 9, 21, 8, 13, 19}

2. {33, 34, 27, 40, 38, 35}

3. {37, 29, 42, 33, 31, 36, 40}

4. {87, 72, 104, 94, 85, 71, 80, 98}

5. {92, 89, 124, 114, 98, 118, 115, 106, 101, 109}

6. {6.7, 3.4, 3.8, 4.2, 5.1, 5.8, 6.0, 4.5}

7. {4.3, 1.9, 6.3, 5.1, 2.1, 1.6, 2.4, 5.6, 5.9, 3.5}

8. {127, 58, 49, 101, 104, 98, 132, 111}

9.

Stem	Leaf
1	0 0 3 8 9
2	0 5
3	1 2 4

2 | 0 = 20

10.

Stem	Leaf
7	8 9
8	1 3 7
9	3 5 6

9 | 3 = 93

11.

Stem	Leaf
0	2 3 6 8 9
1	2 2 5
2	6
3	2 3 4

12.

Stem	Leaf
0	1 1 3 3 7 9
1	2 6 7 8 9 9
2	0 1 2 2 4 5 7 9 9 9
3	2 4 6 7 8
4	0 1 3

2 | 0 = 20

13.

Stem	Leaf
6	0 6
7	1
8	4 9 9
9	1 3 7 7 7 8

8 | 4 = 84

14.

Stem	Leaf
4	8
5	1 2 4 7 7
6	0 2 5
7	4

6 | 2 = 62

HEALTH   For Exercises 15–17, use the data in the table showing the calories burned by a 125-pound person.

15. What is the range of the data?

16. What is the interquartile range of the data?

17. Which activity burns the most calories per hour? The least calories per hour?

Estimated Calories Burned	
Activity	**Calories Burned per Hour**
Basketball	480
Bicycling	600
Hiking	360
Mowing the Lawn	270
Running	660
Soccer	420
Swimming	600
Weight Training	360
Yoga	240

Source: www.fitresource.com

12-3 Skills Practice

Box-and-Whisker Plots

Draw a box-and-whisker plot for each set of data.

1. {6, 9, 22, 17, 14, 11, 18, 28, 19, 21, 16, 15, 12, 3}

2. {$45, $37, $50, $53, $61, $95, $46, $40, $48, $62}

3. {14, 9, 1, 16, 20, 17, 18, 11, 15}

4. {$20, $35, $42, $26, $53, $18, $36, $27, $21, $32}

5. {97, 83, 100, 99, 102, 104, 97, 101, 115, 106, 94, 108, 102, 100, 109, 103, 102, 98, 108}

6. {188, 203, 190, 212, 214, 217, 174, 220, 219, 211, 201, 210, 214, 217, 213, 204, 187, 206, 210}

7.

Goals Scored by MLS Leading Scorers, 2000		
26	15	12
16	15	1
18	11	10
16	15	5
16	13	9

Source: *World Almanac*

8.

Number of 300 Games per Person in Women's International Bowling Congress		
12	17	23
21	17	23
14	21	12
17	19	24
18	27	13
14	20	12
16	12	16

Source: *World Almanac*

12-4 Skills Practice

Histograms

Display each set of data in a histogram.

1.

Shots per Hockey Game		
Number of Shots	**Tally**	**Frequency**
1–7	⊦⊦⊦⊦	5
8–14	I	1
15–21	⊦⊦⊦⊦ III	8
22–28	II	2
29–35	IIII	4

2.

Employees in Each Office		
Number of Employees	**Tally**	**Frequency**
10–19	II	2
20–29	⊦⊦⊦⊦ I	6
30–39	⊦⊦⊦⊦ IIII	9
40–49	⊦⊦⊦⊦ III	8
50–59	I	1

3.

Basketball Backboards on Each Playground		
Number of Backboards	**Tally**	**Frequency**
0–4	⊦⊦⊦⊦ ⊦⊦⊦⊦ ⊦⊦⊦⊦ I	16
5–9	III	3
10–14	⊦⊦⊦⊦ III	8
15–19	⊦⊦⊦⊦ ⊦⊦⊦⊦ I	11
20–24		0
25–29	IIII	4

4.

Population of Loons on Local Lakes		
Number of Loons	**Tally**	**Frequency**
30–39	II	2
40–49		0
50–59	⊦⊦⊦⊦ I	6
60–69	⊦⊦⊦⊦ IIII	9
70–79	⊦⊦⊦⊦ ⊦⊦⊦⊦ ⊦⊦⊦⊦ II	17
80–89	IIII	4

12-5 Skills Practice

Misleading Statistics

For Exercises 1–3, refer to the graphs below.

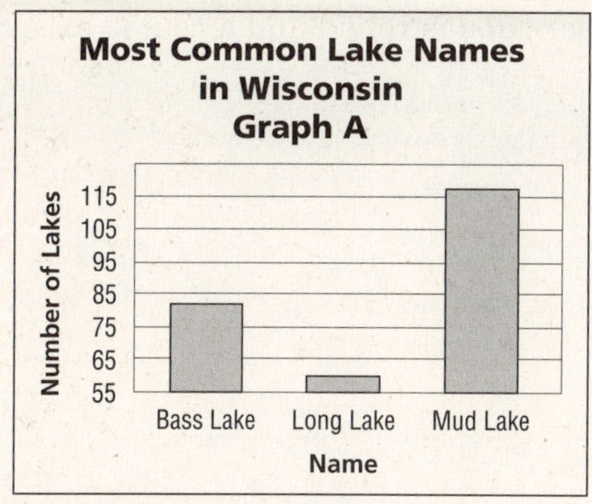

Source: Wisconsin Department of Natural Resources

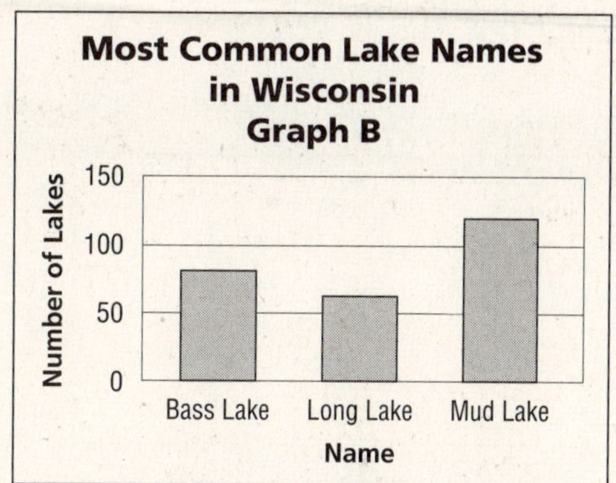

Source: Wisconsin Department of Natural Resources

1. How many lakes in Wisconsin are named Bass Lake? Long Lake? Mud Lake?

2. Which graph gives the impression that only a few lakes are called Long Lake, while numerous lakes are called Bass Lake?

3. What causes the graphs to differ in their appearance?

For Exercises 4–6, refer to the graphs below.

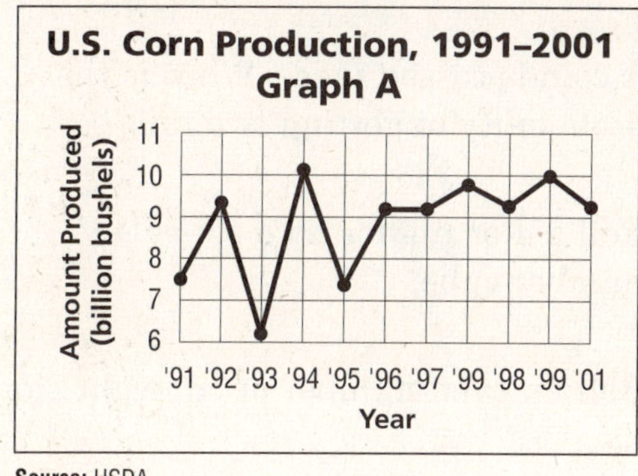

Source: USDA

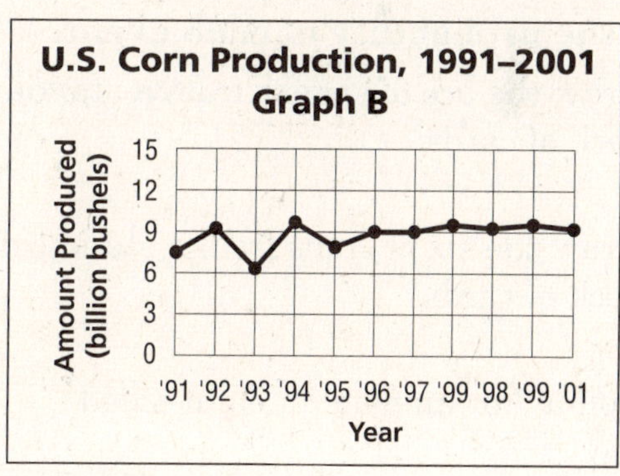

Source: USDA

4. Do these graphs show the same information?

5. Which graph suggests that U.S. corn production is relatively stable?

6. What causes the graphs to differ in their appearance?

12-6 Skills Practice

Counting Outcomes

Draw a tree diagram to find the number of outcomes for each situation.

1. Three coins are tossed.

2. A number cube is rolled and a coin is tossed.

Find the number of possible outcomes for each situation.

3. One card is drawn from a standard deck of cards.

4. Three six-sided number cubes are rolled.

5. One coin is flipped three consecutive times.

6. One coin is flipped and one eight-sided die is rolled.

7. A sweater comes in 3 sizes and 6 colors.

8. A restaurant offers dinners with a choice each of two salads, six entrees, and five desserts.

Find the probability of each event.

9. Draw the ace of spades from a standard deck of cards.

10. A coin is tossed twice. What is the probability of getting two tails?

11. Draw the six of clubs from a standard deck of cards.

12. Roll a 4 or higher on a six-sided number cube.

13. Roll a 7 or an 8 on an eight-sided die.

14. Roll an even number on an eight-sided die.

15. Draw a club from a standard deck of cards.

16. Roll an odd number on a six-sided number cube.

17. A coin is tossed and an eight-sided die is rolled. What is the probability that the coin lands on tails, and the die lands on a 2?

18. A coin is tossed and a card is drawn from a standard deck of cards. What is the probability of landing on tails and choosing a red card?

12-7 Skills Practice

Permutations and Combinations

Tell whether each situation is a *permutation* or *combination*. Then solve.

1. How many ways can 6 student desks be arranged in a row?

2. How many ways can 18 baseball cards be passed out to 2 students?

3. How many ways can 10 students line up for lunch?

4. How many ways can you choose 4 CDs from a stack of 8 CDs?

5. How many ways can 3 pairs of shoes be chosen from 8 pairs?

6. How many ways can 9 runners be arranged on a 4-person relay team?

Find each value.

7. 9!

8. 5!

9. 3!

10. 4!

11. 6!

12. 12!

13. **SPORTS**   The Eastern Division of a baseball league is composed of 5 teams. How many different ways can teams of the Eastern Division finish?

14. **LEISURE**   The local hobby store has 17 model airplanes to display. If the front case holds 6 models, how many ways can 6 planes be chosen for the front of the store?

15. **ZOOS**   The local zoo has 23 animals it can take on visits to schools and other community centers. How many ways can the zoo directors choose 9 animals for a trip to a middle school?

16. **CULTURE**   There are 15 Irish dancers in a championship-level competition. How many ways can the top 3 finishers be arranged?

17. **RACING**   In an auto race, the cars start in 11 rows of 3. How many ways can the front row be made from the field of 33 race cars?

TELEVISION   For Exercises 18 and 19, use the following information.
A television network has a choice of 11 new shows for 4 consecutive time slots.

18. How many ways can four shows be chosen, without considering the age of the viewers or the popularity of the time slots?

19. How many ways can the shows be arranged if the time slots are during prime time and in competition for viewers?

12-8 Skills Practice

Odds

Find the odds of each outcome if an eight-sided die is rolled.

1. three

2. *not* three

3. an even number

4. *not* 7 or 8

5. 1 or 2

6. a number less than 6

Find the odds of each outcome if the spinner at the right is spun.

7. yellow

8. red

9. *not* blue

10. *not* blue or green

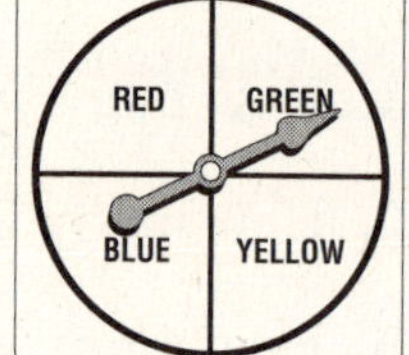

A card is selected from a standard deck of 52 cards.

11. What are the odds of selecting a black jack?

12. What are the odds of selecting a 3?

13. What are the odds of *not* selecting a club?

14. What are the odds of selecting a 6 or an 8?

15. What are the odds of *not* selecting a king?

16. If aces are counted as higher than a king, what are the odds of selecting a card lower than a 10?

BASEBALL   For Exercises 17 and 18, use the following information.

A baseball team finished the season with a record of 92 wins and 70 losses.

17. What were the odds in favor of this team winning any given game during the season?

18. If a player on this team got 150 hits in 500 at bats (a .300 batting average), what were the odds in favor of a hit when he got an at bat?

NUMBER CUBES   For Exercises 19–21, use the following information.

Two number cubes are rolled. The chart at the right shows the outcomes.

19. How many ways can two number cubes total 11? What are the odds of rolling a sum of 11?

20. What are the odds in favor of rolling doubles?

21. What are the odds of *not* rolling a double?

Roll	1	2	3	4	5	6
1	1,1	1,2	1,3	1,4	1,5	1,6
2	2,1	2,2	2,3	2,4	2,5	2,6
3	3,1	3,2	3,3	3,4	3,5	3,6
4	4,1	4,2	4,3	4,4	4,5	4,6
5	5,1	5,2	5,3	5,4	5,5	5,6
6	6,1	6,2	6,3	6,4	6,5	6,6

12-9 Skills Practice

Probability of Compound Events

**A number cube is rolled and the spinner is spun.
Find each probability.**

1. P(2 and green triangle)

2. P(an odd number and a circle)

3. P(a prime number and a quadrilateral)

4. P(a number greater than 4 and a parallelogram)

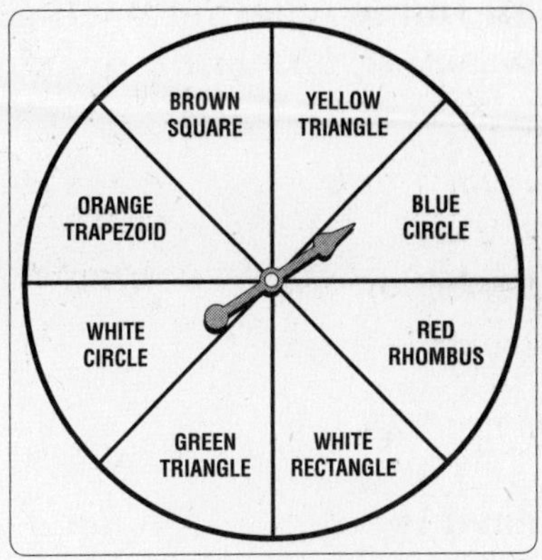

**There are 5 yellow marbles, 1 purple marble, 3 green marbles, and 3 red marbles in
a bag. Once a marble is drawn, it is replaced. Find the probability of each outcome.**

5. a purple then a red marble

6. a red then a green marble

7. two green marbles in a row

8. two red marbles in a row

9. a purple then a green marble

10. a red then a yellow marble

**There are 4 yellow marbles, 3 purple marbles, 1 green marble, and 1 white
marble in a bag. Once a marble is drawn, it is *not* replaced. Find the probability
of each outcome.**

11. a purple then a white marble

12. a white then a green marble

13. two purple marbles in a row

14. two yellow marbles in a row

15. a yellow then a purple marble

16. a green then a white marble

A card is drawn from a standard deck of cards. Find the probability of each outcome.

17. P(a red card or a club)

18. P(a diamond or a spade)

19. P(a face card or a 2)

20. P(a 7 or a 9)

21. P(a red card or a king of spades)

22. P(a heart or a queen of diamonds)

13-1 Skills Practice

Polynomials

Determine whether each expression is a polynomial. If it is, classify it as a *monomial, binomial,* **or** *trinomial.*

1. $-5g^8$

2. $x + 2y + z$

3. $5x + 1 + \dfrac{4}{x}$

4. $r^2 - 9r$

5. $d + 1$

6. $a^3b^2 + a^2$

7. n

8. $17 - \sqrt{c}$

9. $a + b^2 - 3$

10. $m + 2\sqrt{m}$

11. $5y^2 - 3y + 1$

12. $a - b + c$

13. $24x^3$

14. $25 - 9h^4$

15. $u^5 + u^3 + u$

16. $\dfrac{3x^3}{4} + \dfrac{x}{2} + \dfrac{1}{8}$

17. $\dfrac{x}{5} + \dfrac{1}{2}$

18. $\dfrac{6}{a^2} - \dfrac{1}{a} + \dfrac{1}{3}$

19. 1

20. $9y - \sqrt{5}$

21. $27g^5h^2$

Find the degree of each polynomial.

22. 14

23. ab

24. b

25. $c^3 + c^2 + c + 1$

26. mn^5

27. $xy^3z + 1$

28. $k - 4$

29. $\dfrac{-5}{6}$

30. 9.7

31. $c^6de^3 + c^5 + d$

32. $a^2 - 2a + 3$

33. $k^3 + 3k^4$

34. $xy^2 + 4x^2y + y^2$

35. $7b^5 - 10$

36. $16g + 3$

37. $8y^2 + 8y - 5$

38. $abc + 2ab + 5c - bc + 1$

39. $-4g^2h^5 + 2gh^4 + 9$

13-2 Skills Practice

Adding Polynomials

Find each sum.

1. $\begin{aligned} 5q + 7 \\ (+)\ 2q - 2 \end{aligned}$

2. $\begin{aligned} 7f - 10 \\ (+) -2f + 3 \end{aligned}$

3. $\begin{aligned} r^2 - 3r \\ (+)\ r^2 + 4r - 1 \end{aligned}$

4. $\begin{aligned} 9n^2 - 3n \\ (+)\ \ \ \ \ \ 3n - 5 \end{aligned}$

5. $\begin{aligned} w^2 - 3w + 3 \\ (+)\ w^2 + 4w + 1 \end{aligned}$

6. $\begin{aligned} 8c^2 - 4c + 6 \\ (+)\ c^2 + \ c - 1 \end{aligned}$

7. $\begin{aligned} -p^2 + 6p + 8 \\ (+)\ p^2 - 4p - 5 \end{aligned}$

8. $\begin{aligned} 3v^2 + v \\ (+)\ -2v + 7 \end{aligned}$

9. $\begin{aligned} 6m^2 + \ m + 1 \\ (+) 2m^2 - 2m - 3 \end{aligned}$

10. $\begin{aligned} 5d^2 + 7d - 4 \\ (+)\ 5d^2 - 6d - 4 \end{aligned}$

11. $(2r^2 - 3) + (-r^2 + 4r + 1)$

12. $(g^2 + 2g + 5) + (5g^2 - 2g + 3)$

13. $(-m - 9) + (3m - 3)$

14. $(2x^2 + 8x - 7) + (3x + 5)$

15. $(k^2 - k) + (7k^2 - k - 2)$

16. $(4a^2 + 3ab) + (ab + 2b^2)$

17. $(5c - 7) + (3c^2 - 4c + 6)$

18. $(x^2 + xy) + (xy + y^2)$

19. $(-h^2 + 3h - 6) + (4h^2 - 2h + 3)$

20. $(x^2 + x + 1) + (2x - 9x^2)$

21. $(6g^2 - 2g - 3) + (2g^2 + 5g)$

22. $(b^2 + b + 1) + (b^2 - b - 1)$

23. $(2y^2 - 7y + 9) + (y^2 - 4y - 6)$

24. $(7p^3 - 4) + (2p^2 + 5p + 1)$

13-3 Skills Practice

Subtracting Polynomials

Find each difference.

1.
$$\begin{array}{r} 7y + 5 \\ (-)\ y + 6 \\ \hline \end{array}$$

2.
$$\begin{array}{r} k + 8 \\ (-)\ 2k - 9 \\ \hline \end{array}$$

3.
$$\begin{array}{r} w^2 +\ w + 1 \\ (-)\ 2w^2 + 3w + 2 \\ \hline \end{array}$$

4.
$$\begin{array}{r} c^2 - 7c + 2 \\ (-)\ -c^2 -\ c - 1 \\ \hline \end{array}$$

5.
$$\begin{array}{r} 3d^2 -\ d \\ (-)\ d^2 - 3d - 8 \\ \hline \end{array}$$

6.
$$\begin{array}{r} 7n^2 - 3n \\ (-)\ -n^2 - 3n - 1 \\ \hline \end{array}$$

7.
$$\begin{array}{r} 2m^2 - 5m + 3 \\ (-)\ 5m^2 -\ m - 3 \\ \hline \end{array}$$

8.
$$\begin{array}{r} d^2 - 3d - 6 \\ (-)\ d^2 - 2d - 1 \\ \hline \end{array}$$

9.
$$\begin{array}{r} -q^2 + 2q + 2 \\ (-)\ q^2 - 7q + 9 \\ \hline \end{array}$$

10.
$$\begin{array}{r} v^2 +\ v \\ (-)\ 8v^2 - 8v + 8 \\ \hline \end{array}$$

11. $(r^2 - 10r - 3) - (-r^2 - r + 1)$

12. $(7k^2 + k + 8) - (2k^2 - 3k - 3)$

13. $(a^2 - 9) - (a - 4)$

14. $(4x^2 + 11x - 7) - (x^2 - 3x - 6)$

15. $(k^2 - 3k) - (2k^2 - 7k - 1)$

16. $(5a^2 + ab) - (ab + 3b^2)$

17. $(5u^2 - 7) - (3u^2 - 4u + 6)$

18. $(4m^2 + mn) - (3mn + n^2)$

19. $(h^2 + 3h - 6) - (h^2 - 2h - 3)$

20. $(x^2 - x - 1) - (2x + 9x^2)$

21. $(6g^2 + 3g + 3) - (g^2 + g - 5)$

22. $(b^2 + b + 1) - (b^2 - b - 1)$

23. $(a^2 - 9a - 10) - (a^2 - a - 4)$

24. $(4r^2 + 7r) - (3r^2 - 2r + 7)$

13-4 Skills Practice

Multiplying a Polynomial by a Monomial

Find each product.

1. $4(k + 7)$

2. $(5h + 3)3$

3. $-9(2q + 7)$

4. $(6v - 1)(-6)$

5. $-8(5h - 6)$

6. $3(12y - 6)$

7. $(9d + 3)4$

8. $-5(5n - 9)$

9. $2(x^2 + 4)$

10. $-6(5x^2 - 3x)$

11. $(4x^2 - 6x - 9)9$

12. $-7(2c^2 - 8c + 5)$

13. $g(2g + 5)$

14. $-b(9b - 6)$

15. $(4y + 7)y$

16. $(2j - 1)(-j)$

17. $-c(c - 2)$

18. $h(6h + 4)$

19. $(6k + 6)(-k)$

20. $p(3p - 8)$

21. $-a(8a + 2)$

22. $r(r^2 + 7r)$

23. $x(4x^2 - 2x - 1)$

24. $ab(3ab + 2a)$

25. $x(4xy - 3y^2)$

26. $(gh - h)(-g)$

27. $x(4x^2 - xy + y^2)$

28. $6v(3v + 9)$

29. $(u + 4)(-5u)$

30. $8b(b - 6)$

31. $-7d(5d - 9)$

32. $(8w - 4)w$

33. $a(7a + 4)$

34. $(6y - 6)(-y^2)$

35. $s(s + 1)$

36. $-m(6m - 7)$

37. $-k^2(2k - 3)$

38. $c(7c^2 + 3c - 4)$

39. $7mn(m + 2mn + 4n)$

40. $8a(a + ab + b)$

41. $(xy - y^2)(-4xy)$

42. $-8u(7u^2 - 2uv + 4v^2)$

13-5 Skills Practice

Linear and Nonlinear Functions

Determine whether each graph, equation, or table represents a *linear* or *nonlinear* function. Explain.

1. 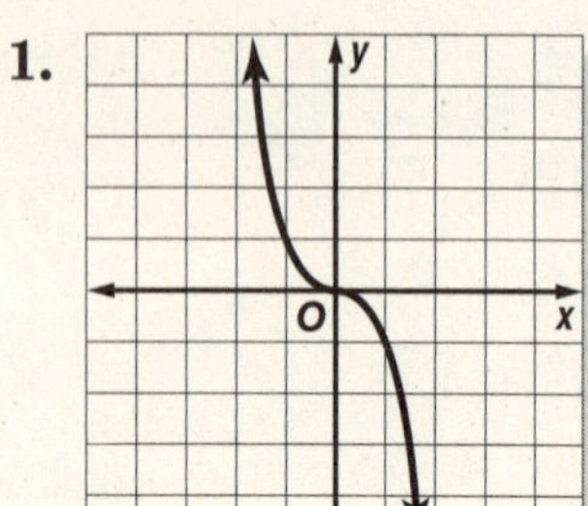

2. 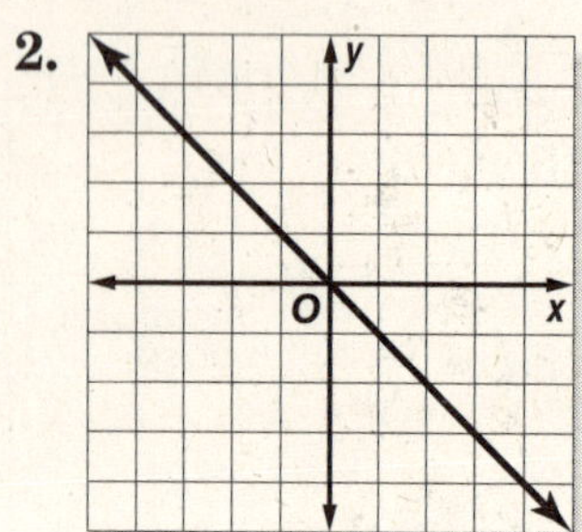

3. 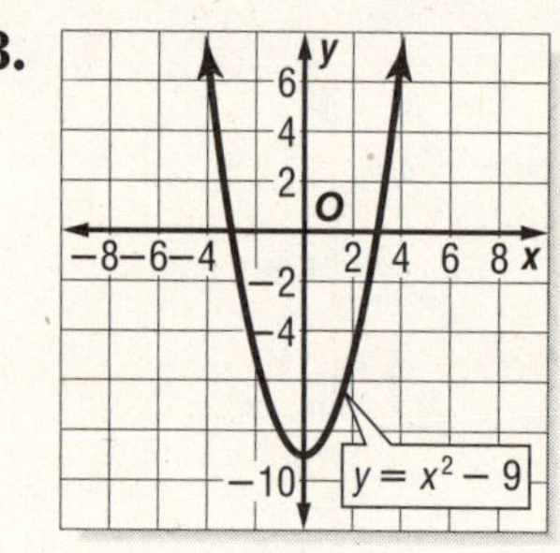

4. $y = \dfrac{x}{2} + 1$

5. $y = \dfrac{2}{x} + 10$

6. $y = 8x$

7. $y = 6$

8. $2x - y = 5$

9. $y = x^2 + 4$

10. $y + 4x^2 - 1 = 0$

11. $2y - 8x + 11 = 0$

12. $y = \sqrt{3x} - 2$

13.

x	y
1	8
2	5
3	2
4	−1

14.

x	y
6	1
12	3
18	6
24	10

15.

x	y
20	−4
15	−2
10	0
5	2

13-6 Skills Practice

Graphing Quadratic and Cubic Functions

Graph each function.

1. $y = 5x^2$

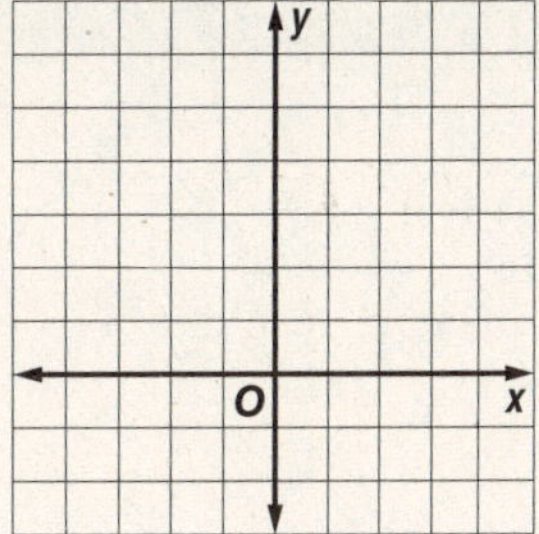

2. $y = 5x^3$

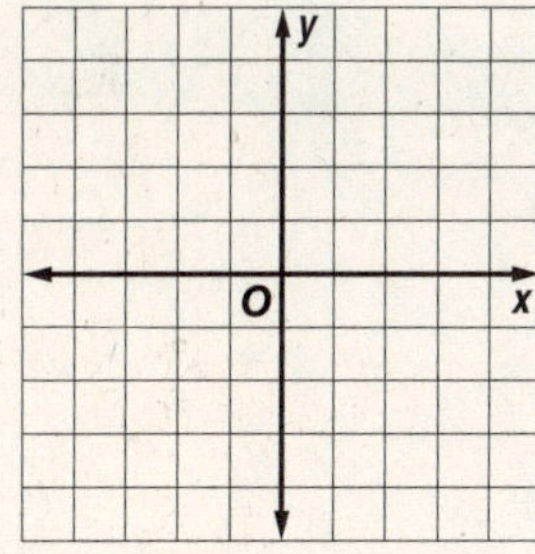

3. $y = -5x^2$

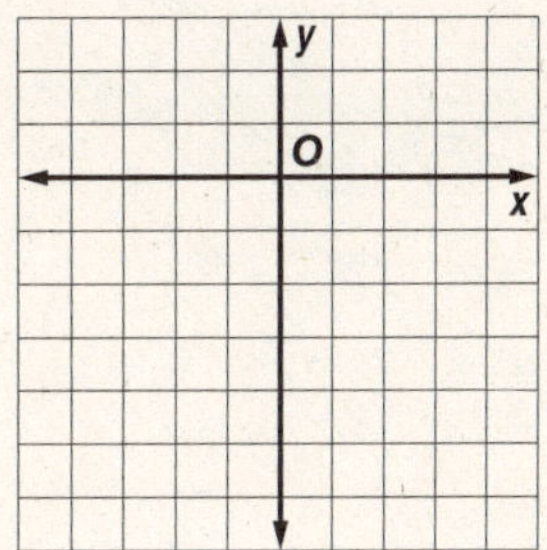

4. $y = -5x^3$

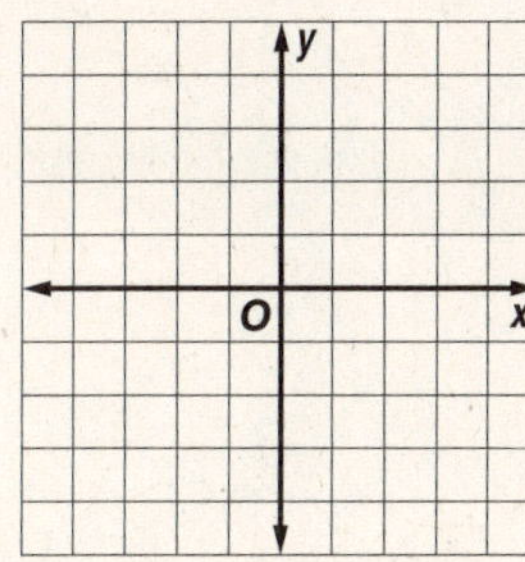

5. $y = x^2 + 4$

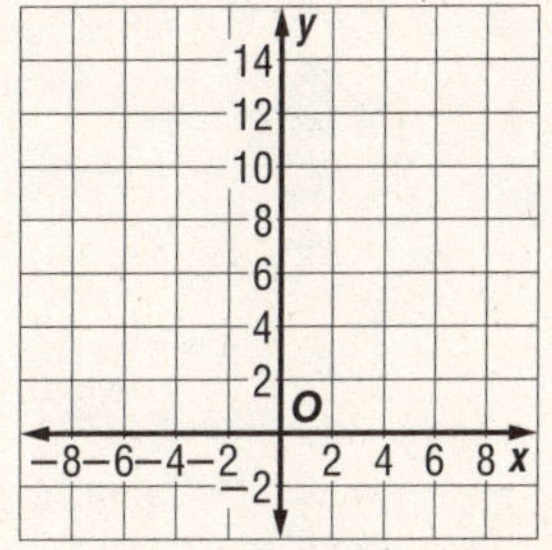

6. $y = x^3 + 4$

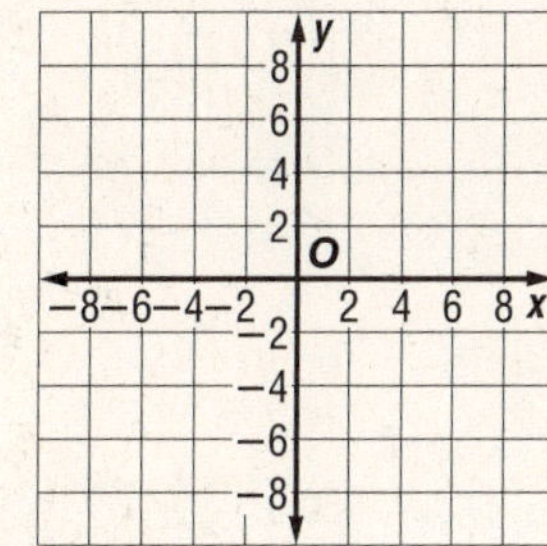

7. $y = x^2 - 4$

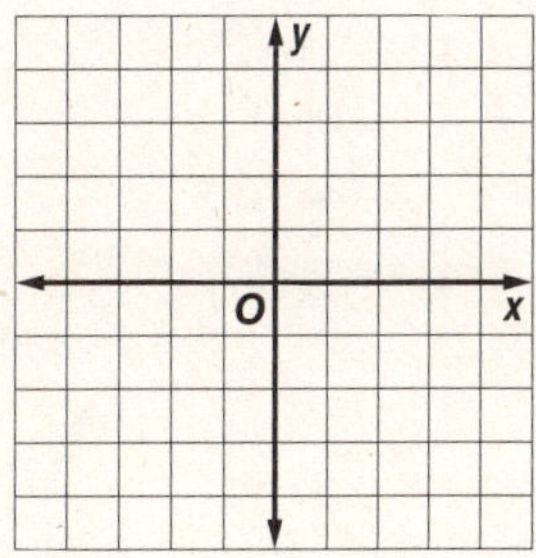

8. $y = x^3 - 4$

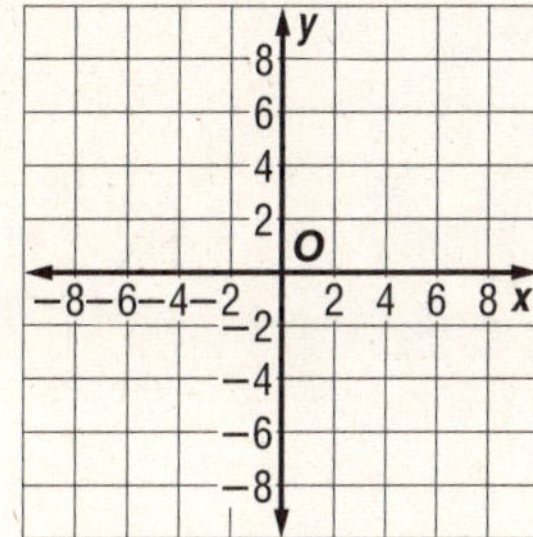

Glencoe Pre-Algebra